荒漠土壤中慢生根瘤菌地理分布及分子进化机制

Geographic Distribution and Molecular Evolutionary Mechanism for *Mesorhizobium* Strains in Desert Soil

冀照君 ◆ 著

哈尔滨

图书在版编目（CIP）数据

荒漠土壤中慢生根瘤菌地理分布及分子进化机制＝Geographic Distribution and Molecular Evolutionary Mechanism for Mesorhizobium Strains in Desert Soil/冀照君著．－－哈尔滨：黑龙江大学出版社，2019.1
ISBN 978-7-5686-0325-6

Ⅰ．①荒… Ⅱ．①冀… Ⅲ．①荒漠－土壤细菌－根瘤菌－地理分布－研究②荒漠－土壤细菌－根瘤菌－分子进化－研究 Ⅳ．①S154.38

中国版本图书馆 CIP 数据核字（2019）第 038498 号

荒漠土壤中慢生根瘤菌地理分布及分子进化机制
Geographic Distribution and Molecular Evolutionary Mechanism for *Mesorhizobium* Strains in Desert Soil
HUANGMO TURANG ZHONG MANSHENG GENLIUJUN DILI FENBU JI FENZI JINHUA JIZHI
冀照君　著

责任编辑	高　媛
出版发行	黑龙江大学出版社
地　　址	哈尔滨市南岗区学府三道街 36 号
印　　刷	哈尔滨市石桥印务有限公司
开　　本	720 毫米×1000 毫米　1/16
印　　张	12.25
字　　数	188 千
版　　次	2019 年 1 月第 1 版
印　　次	2019 年 1 月第 1 次印刷
书　　号	ISBN 978-7-5686-0325-6
定　　价	36.00 元

本书如有印装错误请与本社联系更换。

版权所有　侵权必究

Preface

Mesorhizobium species are the main microsymbionts associated with the medicinal or sand-fixing plants *Astragalus membranaceus* and *Caragana intermedia* (AC) in desert soil in China, while all the *Mesorhizobium* strains isolated from each of these plants could nodulate both of them. Biogeography and molecular evolution of rhizobia influenced by soil environments and selected by legumes have been investigated extensively.

In this book, microevolution of *Mesorhizobium* strains nodulating *Caragana* in semi-fixed desert belt at north China was investigated and investigations on the nodulation of three cultivated medicinal legumes, *Astragalus mongholicus*, *Astragalus membranaceus* and *Hedysarum polybotrys* were performed. Then the whole genomes of two (*M. silamurunense* CCBAU01550, *M. silamurunense* CCBAU45272) and five representative strains (*M. septentrionale* CCBAU01583, *M. amorphae* CCBAU01570, *M. caraganae* CCBAU01502, *M. temperatum* CCBAU01399 and *R. yanglingense* CCBAU01603) originally isolated from AC plants were sequenced, respectively.

Conclusively, the *Caragana*-associated mesorhizobia have divergently evolved according to their geographic distribution, and have been selected not only by the environmental conditions but also by the host plants. The soil fertility may be the main determinants for the distribution of rhizobia associated with these cultured legume plants. The multifactorial features of the rhizobia that may be associated with their host specificity at cross-nodulation group, including *nodE*, *nodZ*, T1SS as the possible main determinants; and *nodO*, hydrogenase system and T3SS as factors regulating the bacteroid formation or nitrogen fixation efficiency.

Zhaojun   Ji

2018. 11

Contents

Chapter I Genetic divergence and gene flow among *Mesorhizobium* strains nodulating the shrub legume *Caragana* growing along the semi-fixed desert belt in north China ······ 1

    Introduction ·· 1

    Materials and methods ·· 2

    Results ·· 16

    Discussion ·· 37

    Conclusion ·· 42

Chapter II Biogeographic distribution of rhizobia associated with the medicinal legumes *Astragalus* spp. and *Hedysarum polybotrys* in agricultural soils ·························· 43

    Introduction ·· 43

    Materials and methods ·· 44

    Results ·· 64

    Discussion ·· 74

Chapter Ⅲ Evolutionarily conserved *nodE*, *nodO*, T1SS, and hydrogenase system in rhizobia of *Astragalus membranaceus* and *Caragana intermedia* 79

Introduction 79

Materials and Methods 80

Results 87

Discussion 102

Conclusion 106

Chapter Ⅳ Genetic divergence among *Bradyrhizobium* strains nodulating wild and cultivated *Kummerowia* spp. in China 107

Introduction 107

Materials and Methods 109

Results 121

Discussion 153

Conclusion 156

References 158

Chapter Ⅰ   Genetic divergence and gene flow among *Mesorhizobium* strains nodulating the shrub legume *Caragana* growing along the semi-fixed desert belt in north China

Introduction

In 14 north provinces of China with an area of 124 043 774 hm^2, deserts, gobi and sand dune have a great influence on the agriculture and livelihood of people and animals dwelling around these regions. Sandstorms are the main natural calamities in these regions especially during the spring. Legumes such as *Caragana*, *Hedysarum scoparium*, *Hedysarum mongolicum*, *Alhagi sparsifolia* and *Sophora alopecuroides* are excellent windproof and sand-fixing plants which widely distribute in arid and semi-arid deserts. Nitrogen fixation of these legumes by establishing symbiosis with rhizobia plays an important role in their healthy growth, especially in the barren sand soil with low content of nutrients. Of the above psammophytic vegetations, *Caragana* distributes more widely and is planted artificially and broadly in the semi-fixed deserts. *Mesorhizobium* spp. , the predominant microsymbionts nodulating the legumes, such as *Caragana* in particular, have a good fitness in alkaline sands due to their capability of acid production, high- or low-temperature and drought resistance. It has been documented that *Mesorhizobium septentrionale*, *Mesorhizobium amorphae*, *Mesorhizobium gobiense*, *Mesorhizobium mediterraneum*, *Mesorhizobium temperatum*, *Mesorhizobium caraganae*, *Mesorhizobium* huakuii, *Mesorhizobium tianshanense*, *Mesorhizobium metallidurans*, and *Mesorhizobium shangrilense* could nodulate *Caragana* in extreme arid and semi-arid deserts, or mountains with low temperature perennially.

So far, the existence of rhizobial biogeographic patterns has been well evidenced, but little is known about the processes underlying them. A central goal of biogeography is to understand the mechanisms that generate and maintain the rhizobial diversity, the natural selection, genetic drift, dispersal and gene mutation on the microevolution of these rhizobia, although a series of interventions on soil stability and preservation of biodiversity have provided tools and strategies to address this problem. Rhizobia nodulating *Caragana* may be used as a model to answer the question because they can survive in stressful environments of the nutrient-poor deserts or inhabit in the root nodule of *Caragana*. Rhizobial genes involved in nodulation and heat-shock play important roles in the symbiosis with legumes and adaptation in stressful environments. Analyses of the genetic divergence and gene flow (the movement and successful establishment of genotypes from one to another) may reveal the adaptively molecular evolution. In order to study the microevolutionary mechanism, six heat-shock factor genes (*clpA*, *clpB*, *dnaK*, *dnaJ*, *grpE*, *hlsU*), five nodulation genes (*nodA*, *nodC*, *nodD*, *nodG*, *nodP*), and five core genes (*atpD*, *glnII*, *gyrB*, *recA*, *rpoB*) were chosen to investigate genetic diversity, gene recombination and mutation events, divergence and gene flow among the *Mesorhizobium* nodulating *Caragana*.

Materials and methods

Rhizobial strains

A total of 724 *Caragana*-nodulating rhizobial strains were selected according to our previous studies. These rhizobial strains originated from 21 sites in 6 areas, including southeast of Tengger Desert (area A, 11 strains), south Mu Us Desert (area B, 447 strains), east Kubuqi Desert (area C, 65 strains), southeast Hunshandake Desert (area D, 72 strains), south Horqin Desert (area E, 113 strains), and mountains of northwest Yunnan (area F, 16 strains). The areas A, B, C, D and E distribute along the northern semi-fixed desert belt in China where the sandstorms blow in spring frequently. All these strains are preserved in the Culture Collection of Beijing Agricultural University (CCBAU) in TY medium with 20% glycerol for a long time.

Of these 724 strains, 72 were selected as representative strains to analyze the molecular evolution (Tab. 1-1). Besides, eight strains including *Mesorhizobium amorphae* ACCC 19665^T, *M. gobiense* CCBAU 83330^T, *M. tianshanense* CCBAU 3306^T, *M. mediterraneum* USDA 3392^T, *M. temperatum* SDW 018^T, *M. metallidurans* LMG 24485^T, *M. huakuii* CCBAU 2609^T and *M. septentrionale* SDW 014^T whose host plants are not *Caragana*, were used as the reference to determine the phylogenetic positions of the *Caragana* mesorhizobia.

In addition, 447 strains obtained from three neighboring sites (B-1, B-2 and B-3) of the area B were analyzed as a subset to study the influence of geographic distance on the gene exchanges among the mesorhizobia.

Tab. 1-1 A total of 72 representative strains isolated from *Caragana* used in this study

Strains (CCBAU No.)	Species	Area	Collector	Province/Autonomous Region	City	Town/District/Banner	Latitude/(°)	Longitude/(°)
75059	*M. amorphae*	A	Li Mao	Ningxia	Wuzhong	Wuzhong	38.127 2	105.917 5
75061	*M. temperatum*	A	Li Mao	Ningxia	Zhongwei	Shaopotou	37.451 1	105.023 0
75063	*M. gobiense*	A	Li Mao	Ningxia	Dujun	Gantang	37.426 1	104.625 8
01570	*M. septentrionale*	B	Li Mao	Inner Mongolia	Ordos	Dongsheng	39.7892	110.129 4
01602	*M. amorphae*	B	Li Mao	Inner Mongolia	Ordos	Dongsheng	39.789 2	110.129 4
01583	*Mesorhizobium* sp. I	B	Li Mao	Inner Mongolia	Ordos	Dongsheng	39.789 2	110.129 4
01577	*Mesorhizobium* sp. VI	B	Li Mao	Inner Mongolia	Ordos	Dongsheng	39.789 2	110.129 4
01582	*M. temperatum*	B	Li Mao	Inner Mongolia	Ordos	Dongsheng	39.789 2	110.129 4
01722	*M. septentrionale*	B	Ji Zhaojun	Inner Mongolia	Ordos	Ejin Horo	39.193 8	109.791 5
01655	*Mesorhizobium* sp. III	B	Ji Zhaojun	Inner Mongolia	Ordos	Ejin Horo	39.193 8	109.791 5
01757	*M. amorphae*	B	Ji Zhaojun	Inner Mongolia	Ordos	Ejin Horo	39.193 8	109.791 5
01731	*M. septentrionale*	B	Ji Zhaojun	Inner Mongolia	Ordos	Ejin Horo	39.193 8	109.791 5
01647	*M. amorphae*	B	Ji Zhaojun	Inner Mongolia	Ordos	Ejin Horo	39.193 8	109.791 5
01790	*Mesorhizobium* sp. II	B	Ji Zhaojun	Inner Mongolia	Ordos	Ejin Horo	39.193 8	109.791 5

Chapter I Genetic divergence and gene flow among *Mesorhizobium* strains nodulating the shrub legume *Caragana* growing along the semi-fixed desert belt in north China

Continued

Strains (CCBAU No.)	Species	Area	Collector	Province/ Autonomous Region	City	Town/District/ Banner	Latitude/(°)	Longitude/(°)
01718	*Mesorhizobium* sp. IV	B	Ji Zhaojun	Inner Mongolia	Ordos	Ejin Horo	39.193 8	109.791 5
01764	*Mesorhizobium* sp. IV	B	Ji Zhaojun	Inner Mongolia	Ordos	Ejin Horo	39.193 8	109.791 5
01800	*Mesorhizobium* sp. VII	B	Ji Zhaojun	Inner Mongolia	Ordos	Ejin Horo	39.193 8	109.791 5
01656	*M. septentrionale*	B	Ji Zhaojun	Inner Mongolia	Ordos	Ejin Horo	39.193 8	109.791 5
01753	*M. septentrionale*	B	Ji Zhaojun	Inner Mongolia	Ordos	Ejin Horo	39.193 8	109.791 5
01645	*Mesorhizobium* sp. VI	B	Ji Zhaojun	Inner Mongolia	Ordos	Ejin Horo	39.193 8	109.791 5
01810	*M. septentrionale*	B	Ji Zhaojun	Inner Mongolia	Ordos	Ejin Horo	39.193 8	109.791 5
01634	*M. septentrionale*	B	Ji Zhaojun	Inner Mongolia	Ordos	Ejin Horo	39.193 8	109.791 5
01662	*Mesorhizobium* sp. III	B	Ji Zhaojun	Inner Mongolia	Ordos	Ejin Horo	39.193 8	109.791 5
01819	*M. mediterraneum*	B	Ji Zhaojun	Inner Mongolia	Ordos	Ejin Horo	39.193 8	109.791 5
01701	*M. septentrionale*	B	Ji Zhaojun	Inner Mongolia	Ordos	Ejin Horo	39.193 8	109.791 5
01660	*M. septentrionale*	B	Ji Zhaojun	Inner Mongolia	Ordos	Ejin Horo	39.193 8	109.791 5
01687	*M. amorphae*	B	Ji Zhaojun	Inner Mongolia	Ordos	Ejin Horo	39.193 8	109.791 5
01669	*M. amorphae*	B	Ji Zhaojun	Inner Mongolia	Ordos	Ejin Horo	39.193 8	109.791 5

Continued

Strains (CCBAU No.)	Species	Area	Collector	Province/ Autonomous Region	City	Town/District/ Banner	Latitude/(°)	Longitude/(°)
01646	M. septentrionale	B	Ji Zhaojun	Inner Mongolia	Ordos	Ejin Horo	39.193 8	109.791 5
01788	M. septentrionale	B	Ji Zhaojun	Inner Mongolia	Ordos	Ejin Horo	39.193 8	109.791 5
01820	M. amorphae	B	Ji Zhaojun	Inner Mongolia	Ordos	Ejin Horo	39.852 2	109.901 1
01821	M. temperatum	B	Ji Zhaojun	Inner Mongolia	Ordos	Ejin Horo	39.852 2	109.901 1
01643	M. septentrionale	B	Ji Zhaojun	Inner Mongolia	Ordos	Ejin Horo	39.852 2	109.901 1
01728	M. septentrionale	B	Ji Zhaojun	Inner Mongolia	Ordos	Ejin Horo	39.852 2	109.901 1
01739	M. septentrionale	B	Ji Zhaojun	Inner Mongolia	Ordos	Ejin Horo	39.852 2	109.901 1
01670	M. septentrionale	B	Ji Zhaojun	Inner Mongolia	Ordos	Ejin Horo	39.852 2	109.901 1
01822	Mesorhizobium sp. XII	B	Ji Zhaojun	Inner Mongolia	Ordos	Ejin Horo	39.852 2	109.901 1
01668	M. amorphae	B	Ji Zhaojun	Inner Mongolia	Ordos	Ejin Horo	39.852 2	109.901 1
01641	M. septentrionale	B	Ji Zhaojun	Inner Mongolia	Ordos	Ejin Horo	39.852 2	109.901 1
01754	M. septentrionale	B	Ji Zhaojun	Inner Mongolia	Ordos	Ejin Horo	39.852 2	109.901 1
01654	Mesorhizobium sp. III	B	Ji Zhaojun	Inner Mongolia	Ordos	Ejin Horo	39.852 2	109.901 1
01751	M. septentrionale	B	Ji Zhaojun	Inner Mongolia	Ordos	Ejin Horo	39.852 2	109.901 1

Chapter Ⅰ Genetic divergence and gene flow among *Mesorhizobium* strains nodulating the shrub legume *Caragana* growing along the semi-fixed desert belt in north China

Continued

Strains (CCBAU No.)	Species	Area	Collector	Province/ Autonomous Region	City	Town/District/ Banner	Latitude/(°)	Longitude/(°)
01752	*M. septentrionale*	B	Ji Zhaojun	Inner Mongolia	Ordos	Ejin Horo	39.852 2	109.901 1
01648	*M. amorphae*	B	Ji Zhaojun	Inner Mongolia	Ordos	Ejin Horo	39.852 2	109.901 1
01661	*M. septentrionale*	B	Ji Zhaojun	Inner Mongolia	Ordos	Ejin Horo	39.852 2	109.901 1
01750	*M. amorphae*	B	Ji Zhaojun	Inner Mongolia	Ordos	Ejin Horo	39.852 2	109.901 1
01636	*M. septentrionale*	B	Ji Zhaojun	Inner Mongolia	Ordos	Ejin Horo	39.852 2	109.901 1
01405	*M. temperatum*	C	Lu Yangli	Inner Mongolia	ulangab	Feng Chin	40.558 3	113.308 7
01597	*M. amorphae*	C	Li Mao	Inner Mongolia	Ordos	Qingshuihe	39.952 8	111.676 7
03299	*M. septentrionale*	C	Lu Yangli	Shanxi	Xinzhou	Pianguan	39.464 6	111.671 9
03254	*Mesorhizobium* sp. Ⅵ	C	Lu Yangli	Shanxi	Xinzhou	Pianguan	39.464 6	111.671 9
01499	*M. caraganae*	D	Lu Yangli	Inner Mongolia	Xilingol League	Duolun	42.196 6	116.498 6
01502	*M. caraganae*	D	Lu Yangli	Inner Mongolia	Chifeng	Balinyou	43.684 8	118.946 1
01477	*M. amorphae*	D	Lu Yangli	Inner Mongolia	Xilingol League	Sanggendalai	42.682 3	115.945 3
01461	*M. loti*	D	Lu Yangli	Inner Mongolia	Xilingol League	Sanggendalai	42.682 3	115.945 3
11185	*M. huakuii*	E	Yan Xuerui	Liaoning	Fuxin	Zhangwu	42.523 8	122.474 2

Continued

Strains (CCBAU No.)	Species	Area	Collector	Province/ Autonomous Region	City	Town/District/ Banner	Latitude/(°)	Longitude/(°)
11196	*M. caraganae*	E	Yan Xuerui	Liaoning	Fuxin	Zhangwu	42.523 8	122.474 2
11206	*Mesorhizobium* sp. V	E	Yan Xuerui	Liaoning	Yingkou	Dashiqiao	40.646 9	122.571 6
11208	*M. amorphae*	E	Yan Xuerui	Liaoning	Yingkou	Dashiqiao	40.646 9	122.571 6
11214	*Mesorhizobium* sp. XI	E	Yan Xuerui	Liaoning	Yingkou	Dashiqiao	40.646 9	122.571 6
11217	*M. caraganae*	E	Yan Xuerui	Liaoning	Yingkou	Dashiqiao	40.646 9	122.571 6
11226	*M. caraganae*	E	Yan Xuerui	Liaoning	Yingkou	Gaizhou	40.668 7	122.233 3
11231	*Mesorhizobium* sp. VIII	E	Yan Xuerui	Liaoning	Yingkou	Gaizhou	40.668 7	122.233 3
11242	*M. amorphae*	E	Yan Xuerui	Liaoning	Chaoyang	Jianping	41.387 6	119.620 3
11244	*M. septentrionale*	E	Yan Xuerui	Liaoning	Chaoyang	Jianping	41.387 6	119.620 3
11257	*M. temperatum*	E	Yan Xuerui	Liaoning	Chaoyang	Jianping	41.387 6	119.620 3
11270	*M. huakuii*	E	Yan Xuerui	Liaoning	Tieling	Kaiyuan	42.285 9	124.225 9
11299^T	*M. caraganae*	E	Yan Xuerui	Liaoning	Chaoyang	Beipiao	41.936 0	121.130 2
65328	*Mesorhizobium* sp. IX	F	Lu Yangli	Yunnan	Diqing	Deqin	28.351 4	99.037 55
65333	*Mesorhizobium* sp. X	F	Lu Yangli	Yunnan	Diqing	Shangri-la	27.907 4	99.831 60

Continued

Strains (CCBAU No.)	Species	Area	Collector	Province/ Autonomous Region	City	Town/District/ Banner	Latitude/(°)	Longitude/(°)
65318	*M. huakuii*	F	Lu Yangli	Yunnan	Diqing	Heqing	26.335 5	100.277 2
65327^T	*M. shangrilense*	F	Lu Yangli	Yunnan	Diqing	Deqin	28.351 4	99.037 55

Note: CCBAU, Culture Collection of Beijing Agricultural University

Gene amplification and sequencing

Template DNA from each strain was extracted as described previously. Five core genes including *atpD*, *glnII*, *gyrB*, *recA* and *rpoB*, encode ATP synthase beta subunit (AtpD), glutamine synthetase II (GlnII), DNA gyrase subunit B (GyrB), recombinase A (RecA), DNA-directed RNA polymerase subunit beta (RpoB), respectively. Six heat-shock factor genes covering *clpA*, *clpB*, *dnaK*, *dnaJ*, *grpE* and *hlsU*, are involved in encoding the ATP-dependent chaperone protein (ClpA and ClpB), molecular chaperone Hsp40 (DnaK and DnaJ), Hsp70 cofactor (GrpE) and ATP-dependent protease (HslU), respectively. Five nodulation genes comprising *nodA*, *nodC*, *nodD*, *nodG* and *nodP*, encode acyltransferase (NodA), N-acetylglucosaminyl transferase (NodC), transcriptional regulator (NodD), 3-oxoacyl-(acyl-carrier-protein) reductase nodulation protein (NodG) and sulfate adenylyltransferase (NodP), respectively.

The PCR amplification protocols of some genes were performed referring the procedures described previously (Tab. 1-2). Primers for the other genes were designed referring to the corresponding homologous regions of the whole genome of *M. amorphae* CCBAU 01578, *M. silamurunense* CCBAU 01550^T, *M. mediterraneum* CCBAU 01399 and *M. caraganae* CCBAU 01502 using the Primer 5.0 software in this study (Tab. 1-2). The annealing temperatures (Tm) for PCR amplification for these heat-shock factors and nodulation genes were also listed in Tab. 1-2. PCR products were purified and commercially sequenced by ABI 3730XL sequencer in Beijing, China, using their corresponding primers. All obtained sequences were checked using ChromasPro (Ver. 1.7.6, Technelysium) and were manually edited using DNAMAN (Ver. 7.212, Lynnon Corp., Quebec, Canada).

Tab. 1-2 Primers designed for the genes and the annealing temperature of them (Tm, ℃) in PCR amplification

Genes	Complete name or function	Preferred primers (5'→3', forward/reverse)	Tm/ ℃	Alternative primers* (5'→3', forward/reverse)	Tm/ ℃	References#
atpD	ATP synthase subunit beta	GCTSGGCCGCATCMTSAACGTC/ GCCGACACTTCMGAACCNGCCTG	58	—	—	—
gln II	Glutamine synthetase II protein	YAAGCTCGAGTACATYTGGCT/ TGCATGCCSGAGCCGTTCCA	58	GCCTTCGAYGGCTCYTC/ CGATGTCGATSYCGTATTT	58	—
gyrB	DNA gyrase subunit B	TTCGACCAGAAYTCCTAYAAGG/ AGCTTGTCCTTSGTCTGCG	57	GTCGGCCGTGTCGGTGGTCA/ TGTCARSCCYTCCCGCCAGT	61	—
recA	Recombinase A	TTCGGCCAAGGMTCGRTSATG/ ACATSACRCCGATCTTCATGC	54	—		—
rpoB	DNA-directed RNA polymerase subunit beta	ATCGTCTCCCAGATGCACCG/ TCGATGTCGTCGATYTCGCC	58	—		—

Continued

Genes	Complete name or function	Preferred primers (5'→3', forward/reverse)	Tm/°C	Alternative primers (5'→3', forward/reverse)	Tm/°C	References[*]
clpA	ATP-dependent Clp protease, ATP-binding subunit protein	GGTYCTYATYCTCAACGA/ CCRACRTAGAGCGGRTT	57	—	—	This study
clpB	ATP-dependent chaperone protein (heat-shock protein)	CGCCCGAACCAAGAACAATCC/ ACCCTCCTCATAGCCGACAT	58	CGTGCCCGAATCGCTGAA/ GGCCTGCCATTTGGTYGTG	60	This study
dnaK	Molecular chaperone DnaK	AAGGAGCAGCAGATCCGCATCCA/ GTACATGGCCTCGCCCGAGCTTCA	62	—	—	—
dnaJ	Chaperone protein DnaJ, heat shock protein (Hsp40), co-chaperone with DnaK	TTCGGCGACATGATGGG/ TTCGGGCABGGVTCCT	58	—	—	This study
grpE	Heat shock protein GrpE (Hsp70 cofactor)	CGGGAAACGGCGATGAG/ VACVTCSGGRTTGG	58	CGGGAAACGGCGATGAG-SGG-/ ARTAGCCBGGCTCACSAC	59	This study

Continued

Genes	Complete name or function	Preferred primers (5'→3', forward/reverse)	Tm/°C	Alternative primers* (5'→3', forward/reverse)	Tm/°C	References#
hslU	ATP-dependent protease ATP-binding subunit HslU	ATCGGYGTCGTYTSAACAT/ CGYTCCATCACSGTCTG	60	—		This study
nodA	Acyltransferase (NodA)	TGCRGTGGARDCTRYGCTGGAAA/ GNCCCTCRTCRAASGTCARGTA	50	—		—
nodC	N-acetylglucosaminyl transferase (NodC)	TGATYGAYATGGARTAYTGGCT/ CGYGACARCCARTCGCTRTTG	52	—		—
nodD	Transcriptional regulator (NodD)	GCCAACGYWTTCTGACACC/ TAAATSCSGGAAGTGGC	55	GCGAACGYWTTCTGACACC/ TCSGTAAATSCSGGAAG	55	This study
nodG	3-oxoacyl-(acyl-carrier-protein) reductase nodulation protein (NodG)	CAAGGACGGYCYCTTCG/ TTCRTYSGARGCGAGRTA	57	RAYCGYGAYGMGGTCAA/ TTCRTYSGARGCGAGRTA	56	This study

Continued

Genes	Complete name or function	Preferred primers (5'→3', forward/reverse)	Tm/ °C	Alternative primers* (5'→3', forward/reverse)	Tm/ °C	References#
nodP	Sulfate adenylyltransferase subunit 2 (NodP)	MTTCAACACYCGCATCG/ ARCCGCAGGTSAGGAAG	58	TCTATTCSGTSGGCAAGG/ ARCCGCAGGTSAGGAAG	57	This study

Note: * The alternative primers were used when the PCR failed using the preferred primers

Primers for some genes were designed referring to the corresponding homologous regions of the whole genome (unpublished) of *M. amorphae* CCBAU 01578, *M. silamurunense* CCBAU 01550^T, *M. mediterraneum* CCBAU 01399 and *M. caraganae* CCBAU 01502 using the Primer 5.0 software in this study

Phylogenetic analyses

The nucleotide sequences of all genes obtained in this study were aligned using the program ClustalW. Neighbor-Joining (NJ) phylogenetic trees were constructed using the MEGA5 program with the Kimura 2-parameter model. Maximum-likelihood (ML) trees were constructed using PhyML software based on the models selected by the Akaike information criterion (AIC) test in the ModelTest 3.7. SplitsTree4 program was used to assess the degree of a tree-like structure for alleles of each locus, and to reveal potentially incompatible signals in the evolutionary history with split phylogenetic networks (1 000 bootstraps).

Nucleotide polymorphism and population genetics analyses

Nucleotide polymorphisms were calculated to count the number of haplotypes (h), haplotype diversity (Hd), nucleotide diversity (π), the π_N/π_S ratios (π_S is the number of synonymous substitutions per synonymous site; π_N is the number of non-synonymous substitutions per non-synonymous site) by DnaSP v5. Population genetic divergence, i.e., average nucleotide divergence (Dxy), and gene flow (or the number of migrants, Nm) were also calculated by DnaSP v5.

ClonalFrame was used to deduce the effect of recombination on the evolution with five independent runs (100 000 burn-in iterations, plus 1 000 000 sampling iterations for each run); satisfactory judgment of which was based on the method described previously. Two recombination rates: r/m (the relative impact of recombination compared with that of the point mutation in the genetic diversification of the lineage), and ρ/θ (the relative frequency of the occurrence of recombination compared with that of point mutation in the history of the lineage) were also calculated.

Admixture levels of the *Mesorhizobium* genospecies were investigated using STRUCTURE with the LOCPRIOR model, and a K value of 4 was chosen for each of the three types of genes with 100 000 burn-in and 100 000 sampling interations. Shimodaira-Hasegawa(SH) test was performed to evaluate the topological consistency among the phylogenetic trees using the PAUP* 4.0b1 program. Minimal recombination events (Rm) within the populations were estimated using the DnaSP v5.

Nucleotide sequence accession numbers

Total 1 137 nucleotide sequences were obtained in this study and were deposited in the GenBank database under accession numbers KP250993 to KP251047, KP251048 to KP251127, KP251128 to KP251207, KP251208 to KP251287, KP251288 to KP251367, KP251368 to KP251425, KP251426 to KP251505, KP251506 to KP251585, KP251586 to KP251665, KP251666 to KP251744, KP251745 to KP251787, KP251788 to KP251850, KP251851 to KP251930, KP251931 to KP252010, KP252011 to KP252053, KP252054 to KP252129, for the genes in order of *atpD*, *clpA*, *clpB*, *dnaJ*, *dnak*, *glnII*, *grpE*, *gyrB*, *hslU*, *nodA*, *nodC*, *nodD*, *nodG*, *nodP*, *recA*, and *rpoB*, respectively.

Results

Phylogenies based on concatenated gene sequences

The phylogenetic tree of the total 724 rhizobial strains were firstly grouped based on *recA* sequences (not shown) and then only 72 strains were selected to represent each of the genotypes for further studies (Tab. 1-1). Topological structures of the phylogenetic trees based on each core gene showed no significant differences using NJ and ML methods (not shown). In addition, the topological structures for the concatenated core (Fig. 1-1) and heat-shock factor genes (Fig. 1-2) were similar in the phylogenetic trees.

Based on the concatenated sequences of the five core genes, 19 genospecies were differentiated according to the ANI threshold value of 96%. Eight genospecies were identified as defined species that corresponded to *M. septentrionale*, *M. amorphae*, *M. gobiense*, *M. mediterraneum*, *M. temperatum*, *M. caraganae*, *M. shangrilense* and *M. huakuii*, respectively. Eleven genospecies showed ANI values lower than 96% with the defined species and they were designated as novel genospecies named *Mesorhizobium* spp. I -XI.

Most of the strains were well grouped and they showed consistency for the topological structures between the core and heat-shock factor gene trees, although some variations could be observed between them, such as CCBAU 01583, CCBAU 01790, CCBAU 01718, CCBAU 01764, CCBAU 01643, CCBAU 01822, CCBAU 01499 and CCBAU 11226 (Fig. 1-2 and Fig. 1-3). However, the phylogenetic relationships of the concatenated sequences of the nodulation genes (Fig. 1-3) were significantly different from those of the core (Fig. 1-1) and heat-shock factor genes (Fig. 1-2).

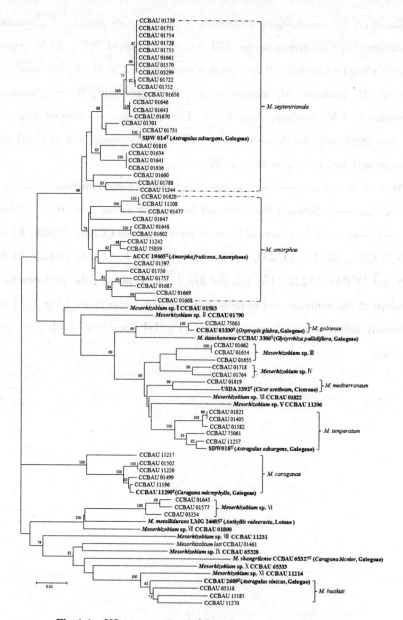

Fig. 1-1 NJ tree constructed based upon the concatenated
sequences of core genes *atpD*, *glnII*, *gyrB*, *recA* and *rpoB*.
Bootstrap values greater than 70% are indicated at the branch points.
The host plants of the type strains are shown in parentheses after the strain numbers.
Novel genospecies (*Mesorhizobium* spp. I -XII) and type strains are boldfaced.
The scale bar represents 1% nucleotide substitutions

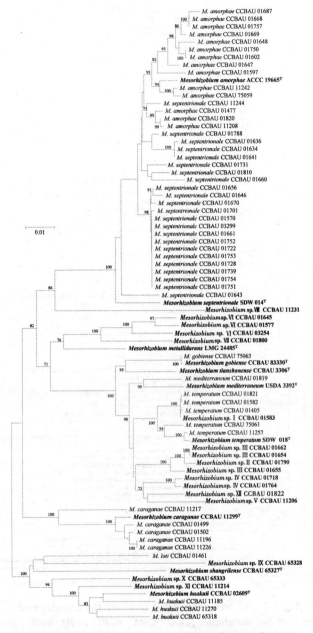

Fig. 1-2 NJ tree constructed based upon the concatenated sequences of heat-shock factor genes. Six genes (*clpA*, *clpB*, *dnaK*, *dnaJ*, *grpE* and *hlsU*) were used. Bootstrap values greater than 70% are indicated at the branch points. Novel genospecies (*Mesorhizobium* spp. I-XII) and type strains are boldfaced. The scale bar represents 1% nucleotide substitutions

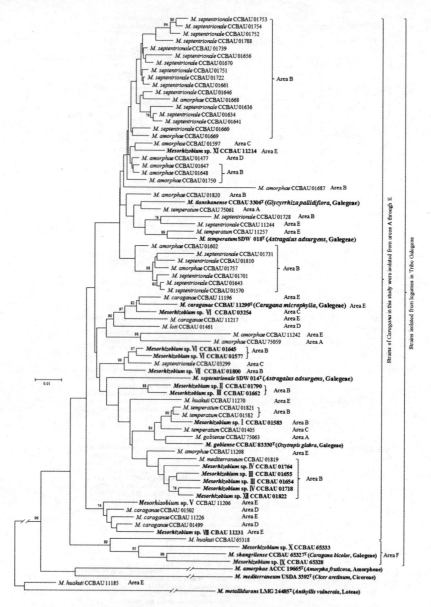

Fig. 1-3 NJ tree constructed based on the concatenated sequences of five nodulation genes (*nodA*, *nodC*, *nodD*, *nodG*, *nodP*), showing the relationships among the *Caragana* (Tribe *Galegeae*). Sootstrap values greater than 70% are indicated at the branch points. The host plants of the type strains are shown in parentheses after the strain numbers. Novel genospecies (*Mesorhizobium* spp. I -XII) and type strains are in bold typeface. The scale bar represents 1% nucleotide substitutions

Nucleotide diversity inferred from different genes

Each of the 72 strains represent a sequence type and the haplotype diversity (Hd) was about 1.000 (Tab. 1-3). The nucleotide polymorphisms for the three kinds of concatenated genes (core, heat-shock factor and nodulation genes) were consistent and the various parameters were listed in Tab. 1-3. The largest deviation was found in nodulation genes, with values of haplotypes (h) from 71 to 25, Hd from 1.0 to 0.805, and the nucleotide diversity (π) from 0.083 75 to 0.026 4. The π values increased gradually in the order of *nodA*, *nodC*, *nodD*, *nodG* and *nodP*, and they were consistent with the variation of π_S values, but not with the π_N values (the highest π_N value was 0.026 6 for *nodD* and the lowest π_N value was 0.010 66 for *nodG*). Strains from all six areas had similar Hd, π, and π_N/π_S values (Tab. 1-4 and Tab. 1-5).

Gene flow and genetic differentiation

NeighborNet network trees for the core, heat-shock factor and nodulation genes were constructed to estimate the gene flow and genetic exchange among strains from different areas. Genetic exchanges were rather common among the strains in areas A-E for the core genes [Fig. 1-4(a)], and for the heat-shock factor genes (Fig. 1-5), but were relatively few for the nodulation genes [Fig. 1-4(b)].

In STRUCTURE analyses, similar structure patterns were obtained for the core and heat-shock factor genes, but they were different from those of the nodulation genes (Fig. 1-6 and Fig. 1-7). Four lineages (I to IV) were clearly identified. As the predominant population, lineage I had exchanged nodulation genes with other lineages much frequently (Fig. 1-6). Gene flow of the core and heat-shock factor genes across the desert belt (area A-E) was significant. By contrast, strains from area F were a lineage that seldom intermingled with the strains from the other five areas for the core and heat-shock factor genes.

As shown in Tab. 1-6, the highest genetic distances (Dxy) were detected and the P values for the Dxy comparison pairs were statistically significant ($P < 0.05$) between area F and the other five areas (mean values of 0.084 62, 0.087 93 and 0.072 78, respectively) for core, heat-shock factor, and nodulation genes. No significant differences were found between the rhizobial populations from areas A, B, C, D and E. However, the mean Dxy values were higher between the populations from areas A to E (0.068 69, 0.069 92, and 0.044 09, for core, heat-shock factor and nodulation genes respectively, $P < 0.05$) than those between the populations from the three sites (B-1, B-2, and B-3) within the same area (0.054 32, 0.055 08 and 0.037 55, for core, heat-shock factor, and nodulation genes, respectively) (Tab. 1-6 and 1-7). Furthermore, the mean Dxy values were higher than the mean π values for all of these areas (Tab. 1-4 and 1-6).

In contrast to area F, where the strains clustered in the fourth branch and evolved from one of the four ancestors, the strains from areas A to E were dispersed across different branches (Fig. 1-6).

Tab. 1-3 Molecular diversity for the three kinds of genes

Genes	Length/bp	S	Eta	h/Hd	π	π_S	π_N	π_N/π_S
Core genes								
atpD	396	99	125	40/0.954	0.053 69	0.207 61	0.008 82	0.042 48
glnII	540	182	233	44/0.951	0.074 32	0.304 40	0.026 13	0.085 84
gyrB	639	221	279	43/0.944	0.074 82	0.292 69	0.023 36	0.079 81
recA	288	75	98	49/0.983	0.055 02	0.234 68	0.001 54	0.006 56
rpoB	591	179	219	46/0.958	0.054 94	0.190 47	0.020 90	0.109 73
Concatenate	2 454	756	954	60/0.989	0.064 19	0.244 47	0.018 52	0.075 76
Average	490.8	151.2	190.8	44.4/0.958	0.062 56	0.245 97	0.016 15	0.064 88
Heat-shock factor genes								
clpA	798	176	217	45/0.966	0.046 09	0.200 21	0.008 25	0.041 21
clpB	597	196	253	45/0.967	0.071 64	0.321 38	0.012 51	0.038 93
dnaK	234	109	147	32/0.876	0.069 72	0.211 19	0.039 03	0.184 81
dnaJ	228	71	103	34/0.928	0.072 13	0.297 42	0.022 26	0.074 84
grpE	366	121	167	38/0.941	0.076 20	0.401 01	0.018 06	0.045 04
hslU	522	184	242	41/0.959	0.067 91	0.246 67	0.022 41	0.090 85

Continued

Genes	Length/bp	S	Eta	h/Hd	π	π_S	π_N	π_N/π_S
Concatenate	2 745	857	1 129	58/0.980	0.065 32	0.265 59	0.016 74	0.063 03
Average	457.5	142.8	188.2	39.2/0.940	0.067 28	0.279 65	0.020 42	0.079 28
Nodulation genes								
nodA	459	98	111	25/0.805	0.026 40	0.070 22	0.012 60	0.179 44
nodC	507	131	149	30/0.912	0.027 18	0.075 95	0.011 85	0.156 02
nodD	492	119	132	71/1.000	0.042 83	0.105 80	0.026 66	0.251 98
nodG	222	66	95	41/0.957	0.065 55	0.288 67	0.010 66	0.036 93
nodP	294	101	142	40/0.942	0.083 75	0.423 00	0.011 54	0.027 28
Concatenate	1974	515	629	79/1.000	0.053 55	0.165 16	0.023 00	0.139 26
Average	394.8	103	125.8	41.4/0.923	0.049 14	0.192 73	0.014 66	0.130 33

Note: S, segregating sites or number of polymorphic (segregating) sites; Eta, total number of mutations; h, haplotype number; Hd, haplotype diversity; π, average number of nucleotide differences per site between two sequences; π_S, nucleotide diversity for synonymous substitutions; π_N, nucleotide diversity for nonsynonymous substitutions

Tab. 1-4　Nucleotide polymorphism six areas (A to F) for the three kinds of genes in *Caragana* mesorhizobia

Area (No. of strains)	Length/bp	S	Eta	h/Hd	π	π_S	π_N	π_N/π_S
Concatenated core genes								
A (3)	2 454	250	261	3/1.000	0.069 41	0.255 46	0.021 69	0.084 91
B (44)	2 454	509	585	35/0.979	0.051 49	0.196 58	0.013 59	0.069 13
C (4)	2 454	310	325	4/1.000	0.071 18	0.277 75	0.019 14	0.068 91
D (4)	2 454	278	294	4/1.000	0.063 23	0.247 77	0.016 29	0.065 75
E (13)	2 454	530	608	13/1.000	0.074 10	0.271 62	0.024 86	0.091 52
F (4)	2 454	313	332	4/1.000	0.071 52	0.284 34	0.017 50	0.061 55
Average	—	365	400.8	10.5/0.997	0.066 82	0.255 59	0.018 85	0.073 63
Concatenated heat-shock factor genes								
A (3)	2 745	286	298	3/1.000	0.070 92	0.300 30	0.014 13	0.047 05
B (44)	2 745	552	638	32/0.958	0.050 49	0.208 13	0.011 44	0.054 97
C (4)	2 745	351	375	4/1.000	0.073 10	0.307 86	0.016 19	0.052 59
D (4)	2 745	331	360	4/1.000	0.068 61	0.274 67	0.018 17	0.066 15
E (13)	2 745	589	697	12/0.987	0.076 33	0.305 83	0.021 04	0.068 80
F (4)	2 745	419	446	4/1.000	0.083 97	0.321 08	0.027 00	0.084 09

Continued

Area (No. of strains)	Length/bp	S	Eta	h/Hd	π	π_S	π_N	π_N/π_S
Average	—	421	469	9.8/0.991	0.070 57	0.286 31	0.018 00	0.062 28
Concatenated nodulation genes								
A (3)	1 974	159	167	3/1.000	0.045 05	0.166 30	0.023 58	0.141 79
B (44)	1 974	341	383	44/1.000	0.036 31	0.116 17	0.012 30	0.105 88
C (4)	1 974	151	165	4/1.000	0.043 23	0.150 16	0.010 94	0.072 86
D (4)	1 974	93	97	4/1.000	0.026 00	0.090 66	0.005 56	0.061 33
E (13)	1 974	365	415	13/1.000	0.010 88	0.155 52	0.025 23	0.162 23
F (4)	1 974	205	220	4/1.000	0.056 23	0.184 37	0.019 53	0.105 93
Average	—	219	241.2	14/1.000	0.036 28	0.143 86	0.016 19	0.108 34

Note: S, segregating sites or number of polymorphic (segregating) sites; Eta, total number of mutations; h, haplotype number; Hd, haplotype diversity; π, average number of nucleotide differences per site between two sequences; π_S, nucleotide diversity for synonymous substitutions; π_N, nucleotide diversity for nonsynonymous substitutions; —, null

Tab. 1-5 Nucleotide polymorphism of three sites within area B for the three kinds of genes

Site (Strain No.)	Length/bp	S	Eta	h/Hd	π	π_S	π_N	π_N/π_S
Concatenated core genes								
B-1 (5)	2 454	325	342	5/1.000	0.066 83	0.257 93	0.018 15	0.070 37
B-2 (22)	2 454	471	534	22/1.000	0.055 66	0.212 32	0.014 86	0.069 99
B-3 (17)	2 454	354	395	13/0.949	0.041 71	0.158 23	0.010 86	0.068 63
Average	—	383.3	423.7	—	0.054 73	0.209 49	0.014 62	0.069 66
Concatenated heat-shock factor genes								
B-1 (5)	2 745	360	380	5/1.000	0.069 76	0.297 46	0.015 16	0.050 96
B-2 (22)	2 745	500	564	21/0.996	0.055 44	0.228 21	0.012 96	0.056 79
B-3 (17)	2 745	368	405	12/0.890	0.037 95	0.153 55	0.008 41	0.054 77
Average	—	409.3	449.7	—	0.054 38	0.226 41	0.012 18	0.054 17
Concatenated nodulation genes								
B-1 (5)	1 974	155	161	5/1.000	0.040 53	0.135 21	0.012 18	0.090 08
B-2 (22)	1 974	300	331	22/1.000	0.039 36	0.126 50	0.013 40	0.105 93
B-3 (17)	1 974	220	236	17/1.000	0.030 73	0.096 36	0.010 69	0.110 94
Average	—	225	242.7	—	0.036 87	0.119 36	0.012 09	0.102 32

Note: S, segregating sites or number of polymorphic (segregating) sites; Eta, total number of mutations; h, haplotype number; Hd, haplotype diversity; π, average number of nucleotide differences per site between two sequences; π_S, nucleotide diversity for synonymous substitutions; π_N, nucleotide diversity for nonsynonymous substitutions; —, null

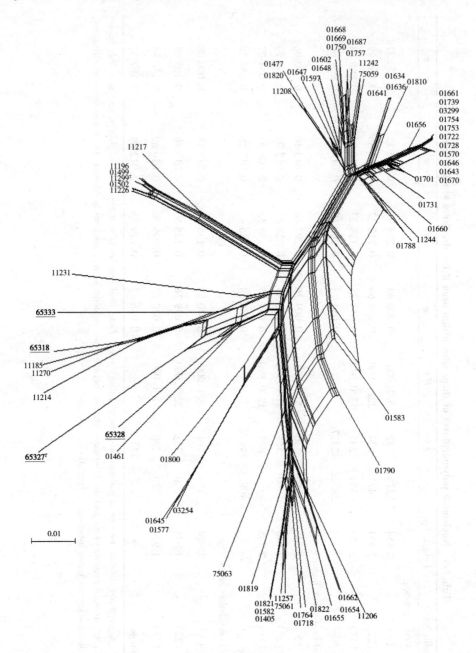

(a)

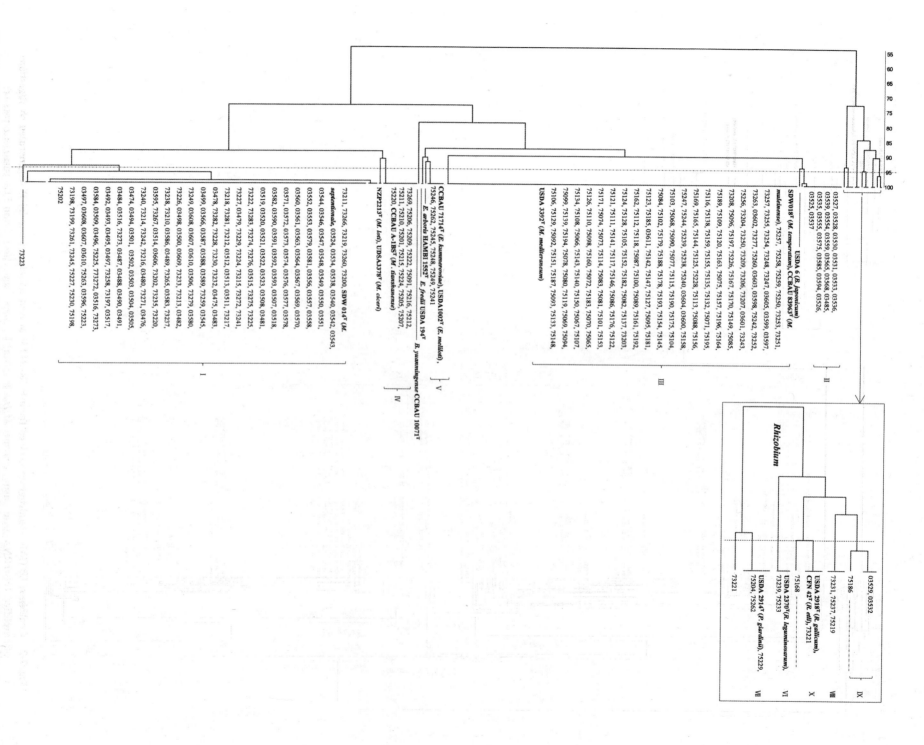

Fig. 2-1 UPGMA dendrogram derived from PCR-RFLP of root nodulate bacterial DNA generated by HaeⅢ, HinfⅠ, MspⅠ and AluⅠ digestion of amplified 16S rRNA gene products. Scale indicates % similarity. The type strain with a superscript (T) after each strain numbers was boldfaced. Some parts of the dendrogram were enlarged to show the strain numbers and branches bigger and clearer. Type strains are boldfaced. E., Ensifer; M., Mesorhizobium; R., Rhizobium; B., Bradyrhizobium

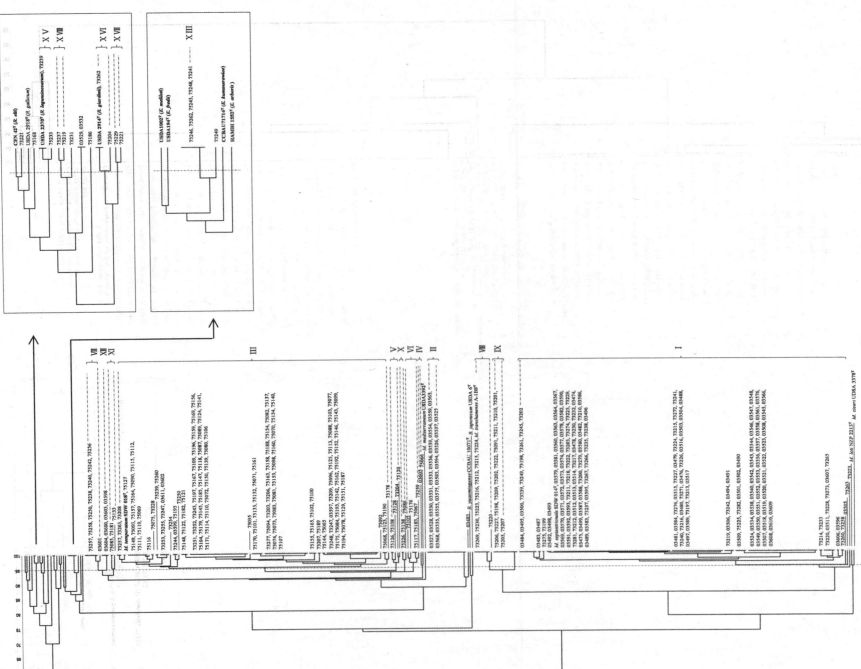

Fig. 2-2 Combined UPGMA dendrogram derived from PCR-RFLP of root nodulate bacterial DNA generated by digestion of amplified 16S rRNA gene (using enzymes of HaeIII, HinfI, MspI and AluI) and ITS (using enzymes of HaeIII, HhaI, MspI) products. The weight ratio of 16S rDNA and IGS was 2:1. Scale indicates % similarity. Some parts of the dendrogram were enlarged to show the strain numbers and branches more bigger and clearer. The type strain with a superscript (T) after each strain number was boldfaced.

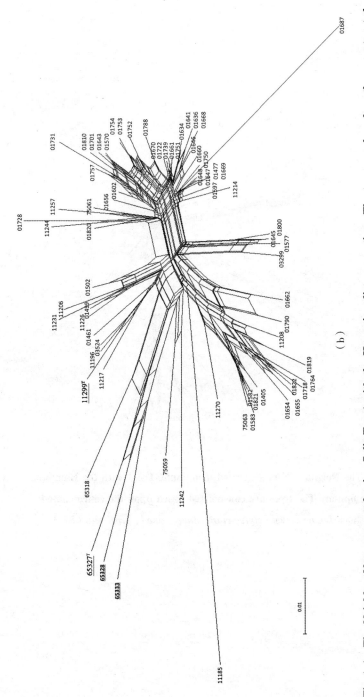

(b)

Fig. 1-4 The Neighbor-Nets generated using SplitsTree4 with the Hamming distance option. The trees are based on the concatenated core gene (a) and nodulation gene sequences (b). Four strains isolated from area F were boldfaced and underlined

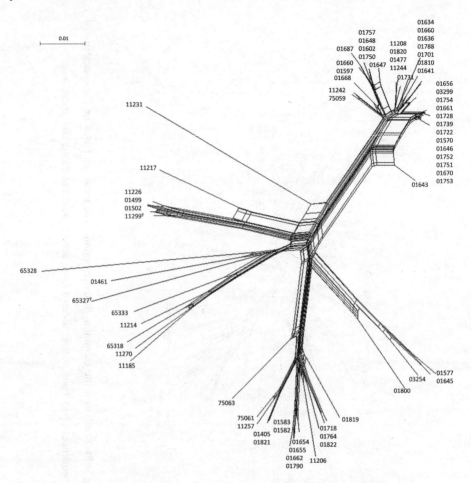

Fig. 1-5 The Neighbor-Nets generated using SplitsTree4 with the Hamming distance option. The trees are constructed based upon the concatenated heat-shock factor genes (*clpA*, *clpB*, *dnaK*, *dnaJ*, *grpE* and *hlsU*)

Chapter I Genetic divergence and gene flow among *Mesorhizobium* strains nodulating the shrub legume *Caragana* growing along the semi-fixed desert belt in north China

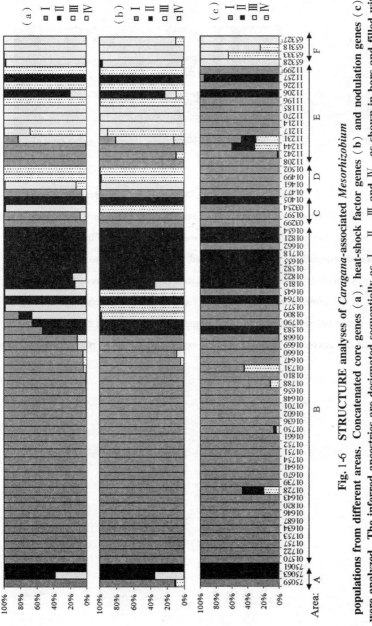

Fig. 1-6 STRUCTURE analyses of *Caragana*-associated *Mesorhizobium* populations from different areas. Concatenated core genes (a), heat-shock factor genes (b) and nodulation genes (c) were analyzed. The inferred ancestries are designated sequentially as Ⅰ, Ⅱ, Ⅲ and Ⅳ, as shown in bars and filled with different gray tones or dots. The horizontal axis represents current *Mesorhizobium* individuals (in the same order in all panels), and the bar for each individual is filled according to the inferred proportions of single-nucleotide alleles that were derived from each of the ancestry. The vertical axis represents the percentage of the strains

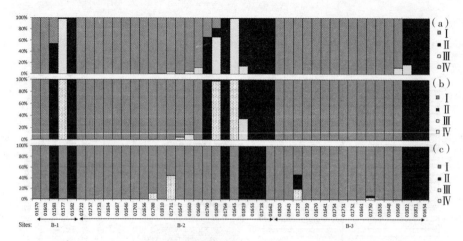

Fig. 1-7 **STRUCTURE** analyses of *Caragana*-associated *Mesorhizobium* populations from three sites within area B. Concatenated core genes (a), heat-shock factor genes (b) and nodulation genes (c) were analyzed. The inferred ancestries are designated sequentially as Ⅰ, Ⅱ, Ⅲ and Ⅳ, as shown in bars and filled with different gray tones or dots. The horizontal axis represents current *Mesorhizobium* individuals (in the same order in all panels), and the bar for each individual is filled according to the inferred proportions of single-nucleotide alleles that were derived from each of the ancestry. The vertical axis represents the percentage of the strains

Tab. 1-6 Genetic divergence (presented as Dxy) and gene flow (presented as Nm) in the *Caragana* mesorhizobia from different areas (A to F)

Dxy	A	B	C	D	E	F
				Nm		
Concatenated core genes						
A	—	13.27	4.66	3.2	14.96	2.15
B	0.062 73^e	—	6.42	2.31	3.42	1.15
C	0.062 75^e	0.056 56^e	—	7.2	42.26	2.25
D	0.076 68^e	0.069 79^e	0.071 87^e	—	19.27	2.5
E	0.074 15^e	0.071 97^e	0.073 50^e	0.066 88^e	—	5.16
F	0.086 86^b	0.088 31^c	0.087 23^b	0.080 84^b	0.079 87^b	—
Concatenated heat-shock factor genes						
A	—	14.57	4.25	4.35	22.75	2.31
B	0.062 79^e	—	6.03	2.2	3.14	1.2
C	0.063 54^e	0.056 67^e	—	11.97	253.93	2.34
D	0.077 78^e	0.073 10^e	0.073 82^e	—	8.56	5.33
E	0.075 24^e	0.073 50^e	0.074 57^e	0.068 24^e	—	6.38
F	0.082 79^b	0.095 3^d	0.095 33^c	0.083 45^b	0.082 79^b	—

Continued

Dxy	Nm					
	A	B	C	D	E	F
Concatenated nod genes						
A	—	194.28	4.69	3.16	19.50	1.67
B	0.046 14[e]	—	72.33	2.27	6.05	0.79
C	0.044 20[e]	0.039 50[e]	—	30.94	11.83	1.17
D	0.047 32[e]	0.038 04[e]	0.035 18[e]	—	15.59	0.76
E	0.053 54[e]	0.048 99[e]	0.046 64[e]	0.041 35[e]	—	1.3
F	0.072 70[b]	0.075 58[d]	0.070 95[b]	0.068 29[b]	0.076 39[d]	—

Note: Number of gene flow (Nm) and average nucleotide divergence between groups (Dxy) are shown in the upper and lower triangular of the Tab.

Statistical difference letter is marked as superscript after each numbers in the lower triangular of the table: b, $0.01 < P < 0.05$; c, $0.001 < P < 0.01$; d, $0.000\ 1 < P < 0.001$; e, non-significant

Tab. 1-7 Genetic divergence (presented as Dxy) and gene flow (presented as Nm) in the *Caragana* mesorhizobia within area B (B-1, B-2, B-3)

Dxy		Nm	
	B-1	B-2	B-3
Concatenated core genes			
B-1	—	12.33	33.03
B-2	0.058 76^e	—	57.86
B-3	0.055 09^e	0.049 11^e	—
Concatenated heat-shock factor genes			
B-1	—	22.3	7.97
B-2	0.061 20^e	—	210.53
B-3	0.057 23^e	0.046 81^e	—
Concatenated nodulation genes			
B-1	—	164.09	8.86
B-2	0.040 07^e	—	165.3
B-3	0.037 64^e	0.034 94^e	—

Note: Number of migrants (Nm) and average nucleotide divergence between groups (Dxy) are shown in the upper and lower triangular of the table.

Statistical difference letter is marked as superscript after each numbers in the lower triangular of the table: b, $0.01 < P < 0.05$; c, $0.001 < P < 0.01$; d, $0.0001 < P < 0.001$; e, non-significant

Recombination of the *Mesorhizobium* lineages in the evolutionary history

As presented in Tab. 1-8, a total of 189, 199, and 150, minimal recombination events (Rm) were inferred for the concatenated core genes, heat-shock factor genes and nodulation genes, respectively.

Results from the recombination in the evolutionary history estimated by the ClonalFrame showed that the relative impact of recombination and mutation (r/m) and the relative frequency of recombination and mutation (ρ/θ) were 0.83 and 0.11 for core genes, 2.10, and 0.54 for heat-shock factor genes, and 0.29, and 0.02 for nodulation genes, respectively (Tab. 1-8). The topologies of all the gene trees, except *rpoB*, were significantly incongruent with the phylogeny of inferred species based on the concatenated sequences of *atpD-glnII-gyrB-recA-rpoB* ($P < 0.05$) in the Shimodaira-Hasegawa (SH) test (Tab. 1-9).

Tab. 1-8 Recombination analysis by DnaSP and ClonalFrame.

Genes	Length/bp	Rm	r/m	ρ/θ
Concatenated core genes	2 454	189	0.83	0.11
Concatenated heat-shock factor genes	2 745	199	2.10	0.54
Concatenated nodulation genes	1 974	150	0.29	0.02

Note: Rm, observed minimum number of recombination events; r/m, the relative impact of recombination compared with that of point mutation in the genetic diversification of the lineage; ρ/θ, the relative frequency of the occurrence of recombination compared with that of point mutation in the history of the lineage

Tab. 1-9 SH test of each gene locus in comparison with the concatenated core genes

Genes	-lnL	Diff-lnL	P
atpD	19 200.85	1 248.80	0.015*
gln II	19 013.43	1 061.38	0.034*
gyrB	18 914.93	962.88	0.041*
recA	20 599.60	2 647.56	0.000*
rpoB	18 583.79	631.74	0.140
clpA	19 303.68	1 351.63	0.008*
clpB	20 563.90	2 611.85	0.000*
dnaJ	20 061.98	2 109.93	0.000*
grpE	19 734.62	1 782.57	0.001*
hslU	19 400.06	1 448.01	0.006*
dnaK	21 942.98	3 990.93	0.000*
nodA	29 462.19	1 1510.14	0.000*
nodC	30 258.78	12 306.73	0.000*
nodD	27 750.07	9 798.02	0.000*
nodG	19 497.57	1 848.87	0.000*
nodP	20 049.94	2 097.89	0.000*

Note: -lnL, negative log-likelihood value for the constrained topology; Diff-lnL, score difference between the nonconstrained and constrained trees; P, significance of the difference in-lnL scores calculated based on the constrained and nonconstrained trees as assessed by SH test; *, $P < 0.05$

Discussion

High genetic diversity of *Caragana*-nodulating rhizobia

The genetic diversity of rhizobia isolated from the root nodules of *Caragana* was very high. Among the 724 rhizobial strains isolated from root nodules of *Cara-*

gana, 19 genospecies were identified (Fig. 1-1) based upon the phylogeny of core genes, which has been introduced to reflect the intraspecific and interspecific difference among the rhizobia previously. These genospecies covered all the previously detected *Caragana*-nodulating *Mesorhizobium* species. In addition, 11 novel genospecies showed a maximum ANI value of 95.8% with the defined species in the concatenated core gene analysis and were designated as *Mesorhizobium* sp. I to XI. Notably high nucleotide polymorphism (Hd and π values) showed that the diversity of these strains was higher than rhizobial strains nodulating soybean.

It was a surprise that the nodulation genes of *Caragana*-nodulating rhizobia were so diverse and no identical genes were detected among the representative strains (Fig. 1-3). However, the rhizobia isolated from *Anthyllis*, *Amorpha*, and *Cicer*, formed long branches deviating from those isolated from *Caragana*, *Astragalus*, *Glycyrrhiza*, and *Oxytropis* (Fig. 1-3), suggesting a consanguineous affiliation of rhizobial nodulation genes interacting and nodulating with the same genus or tribe of legume. Four strains (CCBAU 65318, CCBAU 65333, CCBAU 65327^T, and CCBAU 65328) isolated from area F formed an independent branch, suggesting an effect of geographic isolation on the divergence of the nodulation genes.

Evolutionary relationships among the three types of genes

Rhizobia associated with *Caragana* must possess genes responsible for stress adaptation so that they can survive in the extreme environments of arid and semi-arid deserts. Furthermore, the symbiotic capability of rhizobia is strictly dependent on the nodulation genes. The evolutionary relationships between the heat-shock or stress factor genes, and the nodulation genes are still at unknown. In the present study, we found that the evolutionary history of heat-shock factor genes was very similar to that of the core genes, but was different from the nodulation genes (Fig. 1-1, Fig. 1-2, and Fig. 1-3). The haplotype diversity value (Hd) close to 1.0 for all the three kinds of genes in the 72 strains (Tab. 1-3) means that these strains could fully cover the diversity and well represents the *Caragana* rhizobial populations for further evolutionary analyses. The wider ranges of nucleotide diversity values (π) for nodulation genes than for the core and heat-shock factor genes (Tab. 1-3) supported the hypothesis that the environmental selective pressure for the rhizobial populations were similar. It also means that the level of differentiation in the nodulation genes is greater than that of the other genes. The data in this study also verified that genetic drift has a stronger influence on core and heat-shock factor genes. Moreover, the results suggest that there is a stronger purifying selection on the core genes (average $\pi_N/\pi_S = 0.06488$) and heat-shock factor genes (average $\pi_N/\pi_S = 0.07928$) than on the nodulation genes (average $\pi_N/\pi_S = 0.13033$) (Tab. 1-3). For the nodulation genes, purifying selection was lower for *nodD* gene ($\pi_N/\pi_S = 0.25198$), but stronger for the *nodP* and *nodG* gene ($\pi_N/\pi_S = 0.02728$ and 0.03693, respectively). In addition, the higher π and π_S values, 0.08375 and 0.42300, respectively, inferring from the *nodP* gene suggest that it has a higher synonymous mutation rate than other genes and that it has undergone selection pressures differently from other nodulation genes, such as the conservative *nodA* and *nodC* that are responsible for the synthesis of nodulation factors.

Variation of the nucleotide diversity (π) was not significant for the core and heat-shock factor genes among the rhizobial populations in areas A to E or among the populations in the three sites in area B (Tab. 1-4 and 1-5) suggesting that the selective pressure has no significant influence on the evolutionary history of these strains among these areas.

Gene flow and recombination of the *Caragana* mesorhizobia

Gene flow among the strains isolated from areas A to E for all of the tested genes was more frequent than that among the strains from area F (Fig. 1-4). Genetic divergence was strongly related to the environments due to the higher average nucleotide divergence (Dxy) between the rhizobial groups isolated from the mountains in area F and the other areas (A to E in the desert belt) (Tab. 1-6). In addition, geographic distance also contributed to the higher Dxy values for all of the tested genes, as evidenced by the comparison of the rhizobial populations from areas A to E with those isolated from the three sites within area B (Tab. 1-6 and 1-7).

Horizontal gene transfer (HGT) was observed among the strains isolated from areas A to E and the three sites within area B, as evidenced by the high Nm values and the results of STRUCTURE analyses. Interweave of the three kinds of test genes were clearly found among some strains (Fig. 1-4 and Fig. 1-5).

Gene recombination among different rhizobial populations had been reported previously. Here, extensive recombination was revealed by SH test of the single-gene phylogeny (Tab. 1-9). The incongruences of the phylogenetic trees suggest that the evolutionary directions of these genes are different in the same genospecies. Moreover, the Rm values of 189 and 150, and r/m values of 0.83 and 0.29 further indicate a limited role of recombination in core and nodulation gene evolution. However, for the heat-shock factor genes, the values of Rm was 199, and r/m was 2.10, indicating that it is recombination, but not mutation, plays an important role in the evolutionary history. A strong environmental pressure may result in events of recombination and frequently genetic drift among these *Caragana* mesorhizobia. These results provide multiple lines of evidence that support the natural selection of Darwin's theory and the genetic drift. When mutations could not make a contribution to the rhizobial adaptation and survival, genetic drift may cause these mutations to disappear completely and thereby reduce genetic variation. In addition, the rhizobia tended to acquire some stress-resistant genes by recombination in order to get a better fitness in a specific environment and the recombinants with stress-resistant genes would survive and form new communities by dispersal.

Lineages of the *Caragana* mesorhizobia and adaptive evolution

Four ancestral lineages could be found based upon analyses of the three kinds of genes (Fig. 1-4). The patterns of lineage distribution of the core and heat-shock factor genes were very similar, a phenomenon consistent with results of nucleotide polymorphism (Tab. 1-4). The four ancestral lineages appeared in the desert belt (from areas A to E) and in area F though the imbalance of their distribution (Fig. 1-6 and Fig. 1-7). Lineage I was mainly distributed in areas B, C, and E, while lineage IV was dominant in areas D and E. By contrast, lineage III was found in all of the six areas, although the total number of the strains in this lineage was small. The similar evolutionary history of the core and heat-shock factor genes means that they are essential for survival of *Caragana* mesorhizobia in arid and semi-arid deserts.

Four ancestral lineages were also found for the nodulation genes. Two lineages (I and II) were dominant, which accounted for 98.5% in areas A to E (i. e. the desert belt), while the lineage III was present only in area F (i. e. Yunnan province). Lineage IV was distributed more widely in areas B, E, and F, but the ratio of this lineage was not high (Fig. 1-6). These patterns implied that nodulation genes might be subjected to more strict selection by the host (*Caragana* spp.) in areas A to E, whereas geographic isolation of area F forms the other two lineages (III and IV).

Although the direction of gene mutation could not be determined clearly, the evolutionary history of the core, heat-shock factor and nodulation genes must be restricted by some unknown factors. Natural selection eliminated the individuals whose mutation was disadvantageous to the organisms themselves. Potential selective pressures from physical, chemical and biotic features in soil and antagonistic selection from legumes are the multiple types of stresses to the rhizobia. Genetic divergence within a community may generate from gene flow and genetic drift. However, one rhizobial population surviving successfully from the stressful environment but with a low efficient symbiosis and nitrogen fixation on host plants would not be the dominant one because their symbiotic relationship was weak. This is also true that another rhizobial population with a high efficiency in nitrogen fixation but could not survive in a stressful environment would be dead due to the selective pressure from environments. Genetic diversity of rhizobia in different areas with similar environments are influenced mainly by dispersal and ecological drift and this is may be true for the *Caragana* mesorhizobia in the semi-fixed desert belt in north China.

Conclusion

In conclusion, our results revealed that rhizobial communities were shaped not only by soil environments associating with geographic separation, but also by symbiotic characteristics interacting with their host legumes in the desert belts. The genetic diversity and the evolutionary history of the *Caragana* mesorhizobia were driven mainly by gene mutation and recombination within the rhizobial community, rather than among different areas. The similar evolutionary patterns of the core and heat-shock factor genes of the *Caragana* mesorhizobia in the desert belt may be due to the homogeneous environments.

Chapter II   Biogeographic distribution of rhizobia associated with the medicinal legumes *Astragalus* spp. and *Hedysarum polybotrys* in agricultural soils

Introduction

Compared with natural environments, agricultural cultivation systems provide strong selection stress for both the plants themselves and for the microflora associated with them. Cultivation systems generally provide sufficient light, plentiful water resources, nutrient supplementation in the form of fertilizer and lower competition with other plants. It could be hypothesized that agricultural practices have led to shifts in the associated microflora, especially the rhizobia because the diversity and composition of rhizobial communities are strongly affected by environmental factors, as evidenced by studies on rhizobia of soybean, chickpea and common bean. Rhizobia associated with most cultivated crops, like soybean (*Glycine max*), alfalfa (*Medicago sativa*), pea (*Pisum sativum*), broad bean (*Vicia faba*) and common bean (*Phaseolus vulgaris*), have no natural distribution in natural ecosystems. Therefore, comparative studies on the rhizobial communities associated with the same legume species in natural environments and in agricultural fields are almost absent. Previous studies have been conducted on rhizobia of cultivated species and their wild relative species, such as *Glycine max* and *Glycine soya*, or wild *Vicia* species and cultivated broad bean. Armas-Capote et al. reported that nine *Mesorhizobium* genospecies could nodulate with wild *Cicer canariense* in a natural habitat, while *M. ciceri*, *M. mediterraneum* and *M. muleiense* were the main microsymbionts of chickpea (*Cicer arietinum*) in farmland. However, these studies could not differentiate the rhizobial differences from the same plant species grown in the cultured or wild conditions.

Astragalus mongholicus Bunge and *Astragalus membranaceus* Bunge are nodulating legumes of medicinal importance, which have been used as health-promoting herbs for more than 2 000 years in China. With increasing demand for their production as medicines such as those to treat influenza and to increase immunity, the wild resources of these *Astragalus* species are rather limited. In addition, large scale digging of the roots of these species destroys the vegetation of the ecosystems. Therefore, cultivation of these medicinal legumes has been rapidly and widely developed, especially in the provinces in the northern regions of China. Some rhizobial isolates originating from these two species grown in the wild have been characterized; the major genera isolated were *Mesorhizobium* and *Rhizobium*, with only a few strains belonging to *Sinorhizobium* (now *Ensifer*) and *Bradyrhizobium*. The biogeographic patterns of rhizobial symbionts of *Astragalus* have not been clearly described and the nodulation and diversity of rhizobia associated with cultivated *Astragalus* have not been fully investigated.

The aim of the present study was to investigate the nodulation, diversity and geographic distribution of rhizobia associated with cultured medicinal *Astragalus* species. A small number of rhizobia nodulating with the medicinal legume *Hedysarum polybotrys* were also included because this plant is medicinally similar to *A. membranaceus*. Our current study will provide candidate rhizobial inoculants for these medicinal legumes grown in the agricultural areas.

Materials and methods

Sampling sites and isolation of the nodule bacteria

Root nodules were collected from the cultured plants of *A. mongholicus*, *A.*

membranaceus and *H. polybotrys* grown in 14 sampling sites in the north regions of China (Tab. 2-1). Sites S1 and S2 were in Shanxi Province, N1 through N7 corresponded to Ningxia Province and G1 through G5 were located in Gansu Province. These three provinces were the main cultivation area for the two *Astragalus* species. *A. mongholicus* was abundantly cultured in the three provinces, while *A. membranaceus* and *H. polybotrys* were scatteredly planted in Ningxia and Gansu, respectively. No rhizobial inoculation history was recorded in these sites during the cultivation of these legumes. Specially, site S1 located on a mountain with sandy soil and no fertilizer was ever applied there, while the other sampling sites were in the farming areas with chemical fertilizer used annually. *A. mongholicus* were intercropped with maize only in site S2, while in other sites *A. mongholicus* were grown in a monoculture system.

Root nodules were collected and put into plastic tubes filled with silica gel and were brought to laboratory for further treatments. The shriveled dry nodules were waterlogged to original shape and were surface sterilized using 3% NaClO for 5 min followed by washing 7 times with sterilized water. The bacteria were isolated and purified on yeast mannitol agar (YMA) medium using the standard method. Rhizobial isolates were cultured at 28 ℃ and were storage at 4 ℃ for a short term or in YM broth supplemented with 20% (w/v) glycerol at -80 ℃ for long term reservation.

Tab. 2-1 Isolates used in this study, their genetic types, sampling province, site, and host plant

CC No.	rRNA type	IGS type	Strains (CCBAU No. or specialized by others)	Province/Autonomous Region	Site	Host plant
M. septentrionale						
	I	I	**03524**, 03483, 03474, 03505, 03506, 03484, 03495, 03500, 03487, 03579, 03581, 03560, 03563, 03564, 03567, 03569, 03570, 03571, 03572, 03573, 03574, 03577, 03578, 03582, 03590, 03591, 03592, 03593, 03513, 03494, 03475, 03491, 03492, 03498, 03499, 03587, 03588, 03589, 03580, 03482, 03517, 03586, 03583, 03595, 03584, 03544, 03496, 03489, 03512, 03493, 03497, 03516, 03503, 03504, 03488, 03534, 03538, 03540, 03542, 03543, 03546, 03547, 03548, 03549, 03550, 03551, 03552, 03553, 03556, 03557, 03561, 03576, 03507, 03518, 03519, 03520, 03522, 03523, 03508, 03545, 03566, 03501, 03502, 03481, 03515, 03479, 03480, 03514, 03478, 03476, 03511, 03558, 03521, 03509, 03490	Shanxi	S1	*A. mongholicus*
	I	I	*M. septentrionale* SDW 014^T	Liaoning		*A. adsurgens*

Continued

CC No.	rRNA type	IGS type	Strains (CCBAU No. or specialized by others)	Province/Autonomous Region	Site	Host plant
I	I	I	73218, 73222, 73229, 73212, 73242, 73241, 73235, 73237, 73232, 73219, 73266, 73210, 73238, 73216, 73200, 73211, 73217, 73225, 73230, 73226, 73213, 73227, 73224, 73215, 73240, 73228, 73220, 73233, 73214	Gansu	G1	*H. polybotrys*
I	I	I	73245, 73249, 73261, 73259, 73267, 73260, 73258, 73265	Gansu	G2	*A. mongholicus*
I	I	I	73276, 73275, 73271, 73283, 73272, 73279, 73274, 73281, 73273, 73282	Gansu	G3	*A. mongholicus*
I	I	I	73198, 73199, 73197, 73202	Gansu	G5	*A. mongholicus*
I	I	I	75225	Ningxia	N5	*A. membranaceus*
I	I	II	75263	Ningxia	N5	*A. membranaceus*
M. ciceri						
II	II	III	**03535**, 03559, 03525, 03527, 03528, 03530, 03531, 03533, 03536, 03539, 03554, 03565, 03555, 03575, 03585, 03594, 03526, 03537, 03568	Shanxi	S1	*A. mongholicus*

Continued

CC No.	rRNA type	IGS type	Strains (CCBAU No. or specialized by others)	Province/Autonomous Region	Site	Host plant
M. temperatum						
III		IV	*M. temperatum* SDW 018^T			
			75179, 75151, 75099, 75194, 75192, 75152, 75175, 75193, 75162, 75189, 75188, 75146, 75170, 75178, 75190, 75148, 75182, 75167	Liaoning	—	*A. adsurgens*
III		IV	75185, 75147, 75150, 75171, 75156, 75196, 75159, 75169, 75155, 75165, 75149, 75157, 75164, 75187, 75153, 75154, 75197, 75158, 75163, 75195, 75161, 75160	Ningxia	N2	*A. mongholicus*
III		IV	75119, 75115, 75112, 75111, 75116, 75135, 75109, 75132, 75134, 75127, 75137, 75122, 75121, 75104, 75139, 75106, 75141, 75145, 75118, 75124, 75114, 75110, 75140, 75123, 75144, 75103, 75129, 75125, 75113, 75142, 75143, 75107	Ningxia	N3	*A. mongholicus*
III		IV	75084, 75077, 75090, 75069, 75073, 75083, 75082, 75074, 75070, 75081, 75078, 75088, 75087, 75072, 75080, 75089, 75068, 75071, 75075	Ningxia	N4	*A. mongholicus*

Chapter Ⅱ Biogeographic distribution of rhizobia associated with the medicinal legumes *Astragalus* spp. and *Hedysarum polybotrys* in agricultural soils

Continued

CC No.	rRNA type	IGS type	Strains (CCBAU No. or specialized by others)	Province/Autonomous Region	Site	Host plant
Ⅲ	Ⅲ	Ⅳ	75228	Ningxia	N5	*A. membranaceus*
Ⅲ	Ⅲ	Ⅳ	75093, 75094, 75100, 75098, 75095, 75101, 75102, 75096, 75092	Ningxia	N6	*A. mongholicus*
Ⅲ	Ⅲ	Ⅳ	75065	Ningxia	N7	*A. mongholicus*
Ⅲ	Ⅲ	Ⅳ	75239, 75247, 75260, 75244	Ningxia	N1	*A. mongholicus*
Ⅲ	Ⅲ	Ⅳ	03611, 03602, 03599, 03597	Shanxi	S2	*A. mongholicus*
Ⅲ	Ⅲ	Ⅳ	73243	Gansu	G1	*A. mongholicus*
Ⅲ	Ⅲ	Ⅳ	73251, 73257, 73263, 73252, 73248, 73247	Gansu	G2	*A. mongholicus*
Ⅲ	Ⅲ	Ⅳ	73255, 73253, 73250, 73254	Gansu	G2	*A. mongholicus*
Ⅲ	Ⅲ	Ⅳ	73277	Gansu	G3	*A. mongholicus*
Ⅲ	Ⅲ	Ⅳ	73208, 73207, 73209, 73206	Gansu	G4	*A. mongholicus*
Ⅲ	Ⅲ	Ⅳ	73203	Gansu	G5	*A. mongholicus*
Ⅳ	Ⅲ	Ⅴ	<u>03605</u>	Shanxi	S2	*A. mongholicus*
Ⅵ	Ⅲ	Ⅴ	75066	Ningxia	N7	*A. mongholicus*
Ⅴ	Ⅲ	Ⅵ	<u>73204</u>	Gansu	G4	*A. mongholicus*

Continued

CC No.	rRNA type	IGS type	Strains (CCBAU No. or specialized by others)	Province/Autonomous Region	Site	Host plant
V	III	VI	75120	Ningxia	N1	*A. mongholicus*
V	III	VI	75126, 75108, 75128	Ningxia	N3	*A. mongholicus*
VI	III	VII	<u>75259</u>	Ningxia	N1	*A. mongholicus*
VI	III	VII	75176	Ningxia	N2	*A. mongholicus*
VI	III	VII	75117, 75105, 75183, 75067	Ningxia	N3	*A. mongholicus*
VII	III	I	<u>75238</u>, 75257, 75242, 75256, 75240, 75258, 75250	Ningxia	N1	*A. mongholicus*
VII	III	I	03601	Shanxi	S2	*A. mongholicus*

M. tianshanense

CC No.	rRNA type	IGS type	Strains (CCBAU No. or specialized by others)	Province/Autonomous Region	Site	Host plant
VIII	IV	I	*M. tianshanense* A-1BS^T	Xinjiang	—	*Glycyrrhiza pallidiflora*
VIII	IV	I	<u>75220</u>, 75223, 75230, 75216, 75212, 75224, 75215	Ningxia	N5	*A. membranaceus*
VIII	IV	I	73269	Gansu	G3	*A. mongholicus*
IX	IV	VIII	<u>75206</u>, 75227, 75198, 75209, 75202, 75211, 75210, 75201, 75205, 75207, 75222	Ningxia	N5	*A. membranaceus*
IX	IV	VIII	75091	Ningxia	N4	*A. mongholicus*

Chapter II Biogeographic distribution of rhizobia associated with the medicinal legumes *Astragalus* spp. and *Hedysarum polybotrys* in agricultural soils

Continued

CC No.	rRNA type	IGS type	Strains (CCBAU No. or specialized by others)	Province/Autonomous Region	Site	Host plant
M. muleiense						
X	III	IX	**75138**	Ningxia	N3	*A. membranaceus*
X	III	IX	75086	Ningxia	N4	*A. mongholicus*
X	III	IX	75226	Ningxia	N5	*A. membranaceus*
XI	III	VIII	**75133**	Ningxia	N3	*A. mongholicus*
XI	III	VIII	75085, 75181	Ningxia	N1	*A. mongholicus*
XII	III	X	**03603**, 03604, 03598, 03600	Shanxi	S2	*A. mongholicus*
E. meliloti						
XIII	V	VIII	**75245**, 75246, 75262, 75248, 75241	Ningxia	N1	*A. mongholicus*
XIV	V	XI	**75249**	Ningxia	N1	*A. mongholicus*
R. leguminosarum						
XV	VI	XII	*R. leguminosarum* USDA 2370^T	—	—	*Pisum sativum*
XV	VI	XII	**73239**	Gansu	G1	*H. polybotrys*
XV	VI	XII	75233	Ningxia	N1	*A. mongholicus*

Continued

CC No.	rRNA type	IGS type	Strains (CCBAU No. or specialized by others)	Province/Autonomous Region	Site	Host plant
P. giardinii						
XVI	XII	XIII	*P. giardinii* USDA 2914^T	—	—	*Phaseolus vulagris*
XVI	VII	XIII	73262	Gansu	G2	*A. mongholicus*
XVI	VIII	S	75204	Ningxia	N5	*A. membranaceus*
XVII	S	XI	75221	Ningxia	N5	*A. membranaceus*
XVII	VII	S	75229	Ningxia	N5	*A. membranaceus*
XVIII	XIII	XVI	75219	Ningxia	N5	*A. membranaceus*
XVIII	XIII	XVI	75237	Ningxia	N1	*A. mongholicus*
S	IX	XVII	03529, 03532	Shanxi	S1	*A. mongholicus*
S	II	IV	03485	Shanxi	S1	*A. mongholicus*
S	S	I	73223	Gansu	G1	*H. polybotrys*
S	S	S	75168	Ningxia	N2	*A. mongholicus*
S	VIII	S	73231	Gansu	G1	*H. polybotrys*
S	IX	S	75186	Ningxia	N2	*A. mongholicus*
S	X	S	73221	Gansu	G1	*H. polybotrys*

Chapter Ⅱ Biogeographic distribution of rhizobia associated with the medicinal legumes *Astragalus* spp. and *Hedysarum polybotrys* in agricultural soils

Continued

CC No.	rRNA type	IGS type	Strains (CCBAU No. or specialized by others)	Province/Autonomous Region	Site	Host plant
S	V	S	*E. meliloti* USDA1002^T	—	—	*Meliloti*
S	V	S	*E. kummerowiae* 71714^T	—	—	*Kummerowia stipulacea*
S	S	S	*E. fredii* USDA 194^T	—	—	*Glycine max*
S	X	S	*R. etli* CFN 42^T	—	—	*Phaseolus vulgaris*
S	X	S	*R. gallicum* USDA 2918^T	—	—	*Phaseolus vulgaris*
S	XI	III	*M. loti* NZP 2213^T	—	—	*Lotus corniculatus*
S	XI	S	*M. ciceri* UDSA 3378^T	—	—	*Cicer aretinum*
S	S	S	*E. arboris* HAMBI 1552^T	—	—	*Acacia senegal*
S	S	XVIII	*B. yuanmingense* CCBAU 10071^T	—	—	*Lespedeza*
S	S	XVIII	*B. japonicum* USDA 6^T	—	—	*Glycine max*

Note: * CC, Combined Cluster. S, Single type.
#The information of sampling sites was presented on Tab. 2-2 in details.
Underline: representative strains that only *recA* was sequenced. Underline and bold: Representative strains that *recA*, *atpD*, *gln* Ⅱ, 16S rRNA, *nodC* and *nifH* genes were sequenced.
IGS, intergenic spacer between 16S rRNA and 23S rRNA genes.
The abbreviation *A.* and *H.* stand for *Astragalus*, *Hedysarum*.
The abbreviation *M.* stands for *Mesorhizobium*; *S.*, for *Sinorhizobium* (former *Sinorhizobium*); *R.*, for *Rhizobium*; *B.*, for *Bradyrhizobium*.
—, no need to be shown here. T, type strain. ACCC, Agricultural Culture Collection of China. CCBAU, Culture Collection of Beijing Agricultural University

Tab. 2-2 Locations and altitudes of the sampling sites

Sampling site	Sampling Province (abbr.)	Location#	Longitude (E)	Latitude (N)	Altitude/m
S1	Shanxi (SX)	Zechunling Village, Qianfoling Township, **Hunyuan County**	113°45′30.462″	39°25′57.585″	1 805
S2		**Qin County**	112°36′58.271″	36°59′59.582″	969
N1		Wangleijing Township, **Yanchi County**	106°15′09.129″	37°46′59.535″	1433
N2		Pangzhuang Village, Shenlin Township, **Longde County**	105°58′04″	35°34′57″	1 888
N3	Ningxia (NX)	Shatang Township, **Longde County**	105°51′44″	35°33′37″	1 825
N4		Tianchuan Village, Guanzhuang Township, **Longde County**	106°01′58″	35°44′29″	2 004
N5		Yueyahu Village, Yueyahu Township, **Xingqing District**	106°33′33″	38°37′00″	1 167
N6		Zhaolou Village, Liancai Township, **Longde County**	105°57′44″	35°33′38″	1 840
N7		Xinhe Village, Chenjin Township, **Longde County**	106°08′00″	35°34′28″	2 153

Continued

Sampling site	Sampling Province (abbr.)	Location[#]	Longitude (E)	Latitude (N)	Altitude/m
G1	Gansu (GS)	Xinhua Village, Shouyang Township, **Longxi County**	104°32′10.697″	35°02′36.914″	1 792
G2		Meichuan Township, **Min County**	104°03′54.723″	34°32′45.599″	2 265
G3		Qinxu Township, **Min County**	104°03′22.441″	34°24′14.061″	2 317
G4		Caiziping Village, Shouyang Township, **Longxi County**	104°24′42.556″	35°04′36.277″	1 905
G5		Science Park, **Longxi County**	104°24′46.111″	35°05′19.797″	1 875

Note: [#]The county names in boldface and the sampling provinces (abbr.) were displayed accordingly in the map of China

Characterization of the soil samples

The sample soils were collected from the root zone. After air dried and grinded into powder, the soil samples were sieved through 1.00 mm and 0.15 mm mesh screen successively. Soil samples were delivered to Beijing Academy of Agriculture and Forestry Science for analyses of the pH, total nitrogen (TN), organic carbon (OC), available nitrogen (AN), available phosphorus (AP), available potassium (AK), and electrical conductivity (EC), with the standard methods.

PCR-restriction fragment length polymorphism (RFLP) analysis

Total genomic DNA was extracted from each of the nodule isolates and the reference strains for *Bradyrhizobium*, *Mesorhizobium*, *Ensifer* and *Rhizobium* species using the guanidinium thiocyanate chloride (GuTC) method. Amplification of 16S rRNA genes was carried out using the primers P1 and P6 and the protocol of Tan et al. PCR products (10 μL) were digested separately with each (5U) of the four restriction endonucleases, *Msp*I, *Hinf*I, *Hae*III and *Alu*I at 37 ℃ for 4 h. Digested products were separated by electrophoresis in 2.5% (w/V) agarose gel and stained with 0.5 $\mu g \cdot mL^{-1}$ ethidium bromide. The DNA bands were visualized under UV light and their sizes were estimated by comparing with a 100 bp DNA ladder. As reported previously, similarity of the electrophoretic profiles was estimated by using the Dice coefficient and cluster analysis was performed by using the unweighted pair-group method with arithmetic mean (UPGMA) algorithm and GelCompar II software (version 4.5) (Applied Maths, Sint-Martens-Latem, Belgium). Isolates with identical restriction patterns were classified into the same 16S rRNA type.

The 16S-23S rRNA gene intergenic spacer (ITS) sequences were amplified using the primers FGPL132 and FGPS1490 with the PCR program of Laguerre et al. PCR products were digested with each of the restriction endonucleases *Hae*III, *Msp*I, *Hinf*I and *Hha*I under the same conditions used for the 16S rRNA gene. Separation of the restriction products and analysis of the electrophoretic profiles were as described above. In addition, the restriction fragment length polymorphism (RFLP) patterns of the 16S rRNA gene and ITS (2∶1 weight) sequences were combined in a clustering analysis to obtain a dendrogram for differentiating the genomic species.

Multilocus sequence analysis (MLSA) of different genes

According to the result of the combined RFLP analysis, representative isolates were selected to represent each genomic species to further MLSA of different genes. The following genes were amplified: 16S rRNA gene with the primers P1 and P6; the housekeeping genes *recA*, *glnII* and *atpD* according to the protocols of Vinuesa et al. with the primer pairs recA41F/recA640R, glnII 12F/glnII 689R and atpD255F/atpD782R, respectively; symbiotic gene *nodC* (about 700 bp) using the primer pair nodCF540/nodCR1160 and the protocol of Sarita et al. ; *nifH* gene (784 bp) with the primer pair nifHF/nifHR and the procedure of Laguerre et al. The PCR products were sequenced bi-directionally using the corresponding primers in Beijing Genomics Institute. Together with the related sequences recruited form GenBank database, the obtained sequences in this study were aligned using ClustalW program integrated in MEGA 5.0 and the phylogenetic trees with 1 000 bootstrap replications were constructed using the fitted model and the Maximum Likelihood (ML). Sequences of the three housekeeping genes *recA*, *glnII* and *atpD* were concatenated and a ML phylogenetic tree was constructed. Genospecies were further confirmed by the phylogenetic relationships and by >95% intraspecies sequence similarity value.

Nodulation tests

Representative isolates for different genospecies were cultured separately in 5 mL of YM broth with shaking up to the late exponential phase (about 48 h, $OD_{600} \approx 1.5$) and were inoculated to *A. mongholicus* and *A. membranaceus*, respectively, for testing the nodulation capacity. The seeds were surface sterilized using the method same to the nodule isolation except for 5 min treated in the 3%

NaClO. After rinsing with sterilized water, seeds were incubated 3 days in dark on semisolid agar (0.8%) for germination. The seedlings were sown into the Leonard jars filled with sterilized vermiculite moistured with nitrogen-free nutrient solution. One milliliter of bacterial cultures was inoculated overlying each seedling. Seven replicates were designed for each isolates. The plants were cultured in greenhouse with a cycle of 16 h light at 25 ℃ and 8 h dark at 22 ℃. Nodules were checked and the growth of the plants was recorded after 45 days. The effectiveness of the nitrogen fixation of the nodules was evaluated by the pink color of the nodule section and the green color of the leaves.

Correspondence analysis of rhizobial genospecies and soil factors

Redundancy analysis (RDA), the canonical version of principal component analysis, was used to examine the correspondence between the soil factors and the distribution of rhizobial genospecies in the 14 sampling sites (Tab. 2-3). Community data and soil data (Tab. 2-4) were preanalyzed using WCanoImp and then were analyzed with Detrended Correspondence Analysis (DCA) using Canoco 5.0 (Microcomputer Power, Ithaca, NY). In the DCA, the models of species response to environmental variables and the length of the gradient (first axis) were 0.947, so DCA was proved to be the best method.

Distribution of rhizobia in different soils and plants

Rhizobial diversity, genospecies richness, and evenness in different sampling sites were estimated by three popular alpha ecological indexes: the Shannon-Wiener (H') index calculating genospecies richness in a community; Simpson index (D) and the Pielou index (J), showing the genospecies dominance and evenness, respectively, in a community. These indexes of biodiversity were implemented in the Vegan package (version 2.0-10) and were calculated using the R statistical data language (version 3.1.0). Numbers of the genospecies and isolates from these three legumes was counted and a stacked bar plot was drawn using the barplot function integrated in the R statistical analysis language.

Chapter II Biogeographic distribution of rhizobia associated with the medicinal legumes *Astragalus* spp. and *Hedysarum polybotrys* in agricultural soils

Tab. 2-3 Soil characteristics of sampling sites, distribution and abundance of isolates from nodules of *Astragalus* spp. and *Hedysarum polybotrys*

Site[a]	Soil characteristics[b]							Number of isolates within each genospecies[c]												Total[d]	Index of diversity[e]		
	TN	OC	AN	AP	AK	EC	pH	M1	M2	M3	M4	M5	Si1	R1	R2	R3	R4	R5	R6	R7		H'	D	J
S1	0.5	3.3	15	9.1	31	0.5	7.8	95	—	—	—	20	—	—	2	—	—	—	—	—	117	0.54	0.31	0.49
S2	0.9	9.0	57	8.1	94	0.5	7.8	—	6	4	—	—	—	—	—	—	—	—	—	—	10	0.67	0.48	0.97
N1	0.7	9.3	37	27	106	0.5	7.9	—	13	2	—	—	6	—	—	—	—	1	—	1	23	1.16	0.60	0.72
N2	1.3	30	70	19	211	1.0	8.3	—	41	—	—	—	—	—	—	1	1	—	—	—	43	0.22	0.09	0.20
N3	1.1	25	73	22	220	0.9	8.5	—	39	2	—	—	—	—	—	—	—	—	—	—	41	0.19	0.09	0.28
N4	0.5	5.1	29	11	127	0.8	8.5	—	19	1	1	—	—	—	—	—	—	—	—	—	21	0.38	0.18	0.35
N5	0.8	6.9	51	26	122	2.1	8.2	2	1	1	18	—	—	3	—	—	—	—	—	1	26	1.08	0.50	0.60
N6	1.3	17	71	39	243	1.0	8.4	—	9	—	—	—	—	—	—	—	—	—	—	—	9	0.00	0.00	—
N7	1.2	21	71	31	236	1.0	8.2	—	2	—	—	—	—	—	—	—	—	—	—	—	2	0.00	0.00	—
G1	1.2	12	80	55	170	1.3	7.8	30	1	—	—	—	—	1	—	—	—	1	1	1	34	0.53	0.22	0.33
G2	1.2	14	67	42	280	0.7	7.9	8	10	—	—	—	—	—	—	—	—	—	—	—	19	0.86	0.54	0.78
G3	1.1	9.1	64	17	124	0.8	8.0	10	1	—	1	—	—	—	—	—	—	—	—	—	12	0.57	0.29	0.52

Continued

Site[a]	Soil characteristics[b]							Number of isolates within each genospecies[c]													Total[d]	Index of diversity[e]		
	TN	OC	AN	AP	AK	EC	pH	M1	M2	M3	M4	M5	Si1	R1	R2	R3	R4	R5	R6	R7		H'	D	J
G4	1.1	11	85	30	122	0.9	7.8	—	5	—	—	—	—	—	—	—	—	—	—	—	5	0.00	0.00	—
G5	0.9	8.7	71	12	114	0.7	8.0	4	1	—	—	—	—	—	—	—	—	—	—	—	5	0.50	0.32	0.72
Sum	—	—	—	—	—	—	—	149	148	10	20	20	6	4	2	1	1	2	1	3	367	—	—	—

Note: [a]Sampling sites were described in Tab. 2-2 in details.
[b]TN, total nitrogen (g/kg); OC, organic carbon (g/kg); AN, available nitrogen (mg/kg); AP, available phosphorus (mg/kg); AK, available potassium (mg/kg); EC, electrical conductivity (S/m).
[c]M1, Mesorhizobium septentrionale; M2, M. temperatum; M3, M. muleiense; M4, M. tianshanense; M5, M. ciceri; Si1, Sinorhizobium meliloti; R1, Agrobacterium tumefaciens; R2, Neorhizobium galegae; R3, Rhizobium alamii; R4, Rhizobium sp. I ; R5, R. laguerreae; R6, R. yanglingense; R7, Pararhizobium giardinii.
[d]Total isolates in each sites.
[e]H', Shannon-Wiener's index; D, Simpson's index; J, Pielou's evenness index

Tab. 2-4 Genetic types of RFLP and multilocus sequences of the representative strains from nodules of *Astragalus* spp. and *Hedysarum polybotrys*

Total 25 representatives with CCBAU No. (total isolates)	RFLP type				MLSA[b] similarity to defined species
	16S	IGS	Comb.	Cluster[a]	
Mesorhizobium septentrionale					
03524 (148)	I	I		I	96.0-97.5% to *M. septentrionale* SDW 014[T]
73223 (1)	Single	I		Single	
M. temperatum					
75179 (127)	III	IV		III	97.2-99.6% to *M. temperatum* SDW 018[T]
75238 (8)	III	I		VII	
73204 (5)	III	VI		V	
03605 (2)	III	V		IV	
75259 (6)	III	VII		VI	
M. muleiense					
75138 (3)	III	IX		X	96.1-98.7% to *M. muleiense* CCBAU 83963[T]
75133 (3)	III	VIII		XI	
03603 (4)	III	X		XII	
M. tianshanense					
75206 (12)	IV	VIII		IX	97.3-97.9% to *M. tianshanense* A-1BS[T]
75220 (8)	IV	I		VIII	

Continued

Total 25 representatives with CCBAU No. (total isolates)	RFLP type			MLSA[b] similarity to defined species
	16S	IGS	Comb. Cluster[a]	
M. ciceri				
03535 (19)	II	III	II	96.7-97.2% to *M. ciceri* USDA 3383[T]
03485 (1)	II	IV	Single	
Sinorhizobium meliloti				
75245 (5)	V	VIII	XIII	98.2-98.4% to *S. meliloti* USDA 1002[T]
75249 (1)	V	XI	XIV	
Agrobacterium tumefaciens				
75221 (2)	Single	XI	XVII	99.1-100% to *A. tumefaciens* B6[T]
73262 (2)	VII	XIII	XVI	
Neorhizobium galegae				
03529 (2)	IX	XVII	Single	95.2% to *N. galegae* LMG 6214[T]
Rhizobium alamii				
75168 (1)	Single	Single	Single	97.3% to *R. alamii* GVB 016[T]
Rhizobium sp. I				
75186 (1)	IX	Single	Single	88.9% to *R. cellulosilyticus* LMG 23642[T]
R. leguminosarum				
73239 (2)	VI	XII	XV	96.3-96.6% to *R. leguminosarum* USDA 2370[T]

Continued

Total 25 representatives with CCBAU No. (total isolates)	RFLP type			MLSA[b] similarity to defined species
	16S	IGS	Comb. Cluster[a]	
R. yanglingense				
73221 (1)	X	Single	Single	99.9% to *R. yanglingense* SH22623[T]
Pararhizobium giardinii				
75219 (2)	VIII	XVI	XVIII	99.0-99.1% to *P. giardinii* R-4395[T]
73231 (1)	VIII	Single	Single	

Note: [a]Comb. Cluster, combined cluster based upon the combination of 16S rRNA and IGS gene PCR-RFLP.
[b]MLSA, multilocus sequence typing analyses based on the combined gene sequences of *recA*, *atpD* and *glnII*

Results

Nodulation of the plants and isolation of root nodule bacteria

Nodulation of the annual plant A. *membranaceus* in the agricultural fields was relatively poor and only a few nodules were collected from this species. Claviform nodules were easily obtained from the taproot or lateral roots of 1- or 2-year-old individuals of perennial A. *mongholicus* and H. *polybotrys* plants, but it was difficult to find nodules on the roots of plants more than 3 years old. H. *polybotrys* was only planted in Gansu Province, where the other *Astragalus* species were cultured widely in Shanxi, Gansu and Ningxia. In total, 367 isolates were obtained from the root nodules of the three medicinal legumes. The geographic origins of these isolates were: 127 from nodules of the two sites (S1 and S2) in Shanxi, 165 isolates from nodules of the seven sites (N1 through N7) in Ningxia and 75 isolates from nodules of the five sites (G1 through G5) in Gansu (Tab. 2-3).

Chemical properties of the soils

The physical and chemical features of the sampled soils were listed in Tab. 2-3. All of the sampling sites presented weakly alkaline soil with pH ranging from 7.8 to 8.5. The contents (mg/kg) of available nitrogen (AN), available phosphorus (AP) and available potassium (AK) were 15-85, 8.1-42 and 31-280, respectively. The electrical conductivity (EC) values (S/m) varied from 0.5 in sites S1, S2 and N1 to 2.1 in site N5. Contents of total nitrogen (TN) fluctuated from 0.5 in sites S1 and N4 to 1.3 in sites N2 and N6. Sampling site S1 had the most barren soil and the lowest values for various soil parameters, including TN, organic carbon (OC), AN, AK, AP, EC and pH (Tab. 2-3).

RFLP of the 16S rRNA gene and ITS sequence PCR products

In this study, 13 rRNA types and 21 ITS types were identified among the 367 isolates in the RFLP analyses (Tab. 2-4). The analysis of combined PCR-RFLP of 16S rRNA gene and ITS sequence classified the isolates into 25 clusters at the similarity level of 95%, in which 7 clusters contained only single isolates (Tab. 2-4 and Fig. 2-1). Combined clusters I and III were the dominant groups containing 148 and 127 isolates, respectively and they were closely related to *M. septentrionale* and *M. temperatum*, respectively (Fig. 2-2). The other minor clusters contained 1 to 19 isolates grouping with the reference strains *M. tianshanense*, *M. ciceri*, *Ensifer meliloti*, *Agrobacterium tumefaciens*, *Rhizobium leguminosarum* and *Pararhizobium giardinii* or existing as single clusters (Tab. 2-4).

Sequence analyses of 16S rRNA and housekeeping genes

In this analysis, 25 isolates were selected to represent the major combined clusters obtained by RFLP analysis (Tab. 2-4). Fourteen isolates were grouped with the *Mesorhizobium* reference strains and they were related to *M. muleiense*, *M. temperatum*, *M. tianshanense*, *M. septentrionale* and *M. ciceri* (Tab. 2-4 and Fig. 2-3). Two representative isolates were close to *E. meliloti* (and *E. kummerowiae*). The other isolates were related to *N. galegae*, *P. giardinii*, *R. yanglingense*, *R. alamii*, *R. laguerreae*, *A. tumefaciens* and a novel lineage (CCBAU 75186) (Tab. 2-4 and Fig. 2-3).

Twenty-seven representative isolates were classified into 11 defined species and one unknown *Rhizobium* sp. based upon sequence similarities of 95% (Tab. 2-4) of the concatenated *recA*, *glnII* and *atpD* gene sequences and the MLSA phylogenetic tree (Fig. 2-4). Five genospecies belonged to the genus *Mesorhizobium* corresponding to *M. septentrionale* (149 isolates), *M. temperatum* (148 isolates), *M. muleiense* (10 isolates), *M. tianshanense* (20 isolates) and *M. ciceri*

(20 isolates) (Tab. 2-4). These mesorhizobia accounted for 94.5% (347 isolates) of the total isolates. One genospecies (including 6 isolates, combined cluster XIII and XIV) in *Ensifer* was classified as *E. meliloti*. Four representative isolates were identified as *Agrobacterium tumefaciens*. A single isolate (CCBAU 75186) was identified as *Rhizobium* sp. because of its low similarity (88.9%) to *R. cellulosilyticum* LMG 23642^T. Nine genospecies (containing 1 to 3 isolates) were identified as *N. galegae*, *R. alamii*, *R. laguerreae*, *R. yanglingense* and *P. giardinii* (Tab. 2-4).

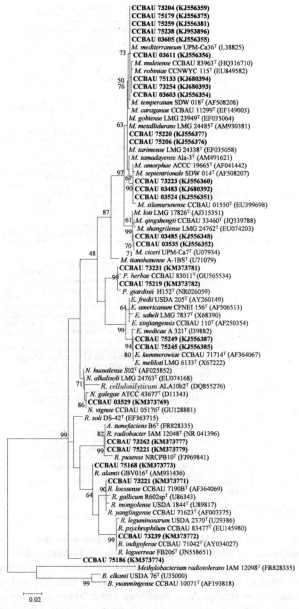

Fig. 2-3 Maximum likelihood phylogenetic tree of 16S rRNA gene sequences. T92 + G was used as the model. Isolates isolated from medical legumes in this study are shown in bold. Bootstrap values ≥50% are given at the branching points. Bar, 2% nucleotide substitution per site. Superscript (T) after strain number indicates the type strain of the species. Accession numbers of sequences for 16S rRNA genes extracted from GenBank are given in the parentheses

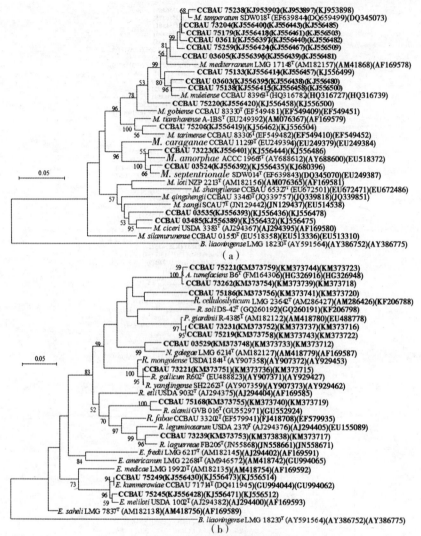

Fig. 2-4 ML phylogenetic trees based on the concatenated sequences of *recA*, *atpD* and *gln* Ⅱ genes. Isolates isolated from medical legumes in this study are shown in bold. These ML trees were constructed with T92 + G + I model for *Mesorhizobium* species (a) or with T92 + G model for *Ensifer* and *Rhizobium* species (b). Bootstrap values above 50% are given at the branching points. Bars, 5% nucleotide substitution per site. Superscript (^{T}) after strain number indicates the type strain of the species. Accession numbers of sequences for *recA*, *atpD* and *glnⅡ* genes extracted from GenBank are given in parentheses. *Bradyrhizobium liaoningense* LMG 18230^T was used as an outgroup in each tree

Analysis of symbiotic genes and the nodulation test

The symbiotic genes *nodC* and *nifH* could only be amplified from the representative isolates in genera *Mesorhizobium* and *Ensifer*. Most of the isolates in *Mesorhizobium* had type I *nodC* while only one isolate (CCBAU 73223) had type II *nodC* [Fig. 2-5 (a)]. Two isolates (CCBAU 75249 and CCBAU 75245) possessed *nodC* genes with high similarities to *E. meliloti* (type III *nodC*) [Fig. 2-5(b)]. The topological structure of the *nifH* phylogenetic tree (Fig. 2-6) was similar to that of *nodC* (Fig. 2-5) but only two *nifH* types, I and II, corresponding to *nodC* types I and III, respectively, were found. In the nodulation test, all the isolates belonging to *Mesorhizobium* could form effective nodules with *A. membranaceus* and *A. mongholicus*. Isolates from other genera could not nodulate with these two medicinal legumes.

Geographic distribution of the rhizobia and their ratio in each legume species

As described above, *Mesorhizobium* species were the predominant microsymbionts of the three medicinal legumes grown in different sampling sites (Tab. 2-3). Of these genospecies, *M. temperatum*, *M. muleiense* and *M. septentrionale* were the three most widely distributed rhizobial genospecies and they could be found in nodules of 13, 5 and 6 sampling sites, respectively (Tab. 2-3). *M. tianshanense* mainly distributed in nodules of the sampling sites G3, N5 and N4, while *M. ciceri* was only isolated from nodules in site S1 (Tab. 2-3).

The diversity index (Tab. 2-3) of Shannon-Wiener (H') in nodules in site N1 was the highest (1.16), followed by that in site G5 (1.08) and site G2 (0.86). The lowest H' value (0) was found in nodules in sites N7, N6 and G4 because only one genospecies was isolated from the nodules there. The remaining sampling sites had H' values between 0.67 and 0.22. The values of Simpson's index (D) varied between 0 and 0.60 in nodules of the 14 sampling sites and were consistent with the H' values. Pielou's evenness index (J) varied from 0.20 for site N2 to 0.97 for site S2. These values indicated that the diversity and genospecies composition of the *Astragalus* rhizobial community varied dramatically in nodules in different sampling sites (Tab. 2-3).

Among the 367 isolates, 307 were from *A. mongholicus*, 27 from *A. membranaceus* and 33 from *H. polybotrys*. *M. septentrionale*, *M. temperatum* and *M. ciceri* were the major rhizobial species isolated from *A. mongholicus*, while *M. tianshanense* and *M. septentrionale* were the major microsymbionts from *A. membranaceus* and *H. polybotrys*, respectively (Fig. 2-7).

Correlation between the distribution of rhizobia and soil characteristics

According to the lengths of the arrows, the angles between the arrows and the distributions of rhizobial genospecies shown in Fig. 2-8, OC, pH and AN were the most important soil factors influencing the distribution of rhizobia. 3 genospecies, *M. septentrionale*, *M. ciceri* and *N. galegae*, distributed mainly in nodules of the sampling site S1 and their distributions were negatively correlated with soil TN, AN, AP, pH and OC contents. In contrast, *R. yanglingense* and *P. giardinii* tended to be associated with legumes grown in soils with low OC and pH values such as in sampling site S2. The pH was positively correlated with the distribution of *M. temperatum*, *M. muleiense*, *R. alamii* and *Rhizobium* sp. I in nodules collected from sampling site N2, but was negatively correlated with the distribution of other genospecies (Fig. 2-8). EC and AP were negatively correlated with the distribution of *M. muleiense* but positively correlated with the distribution of *A. tumefaciens* and *M. tianshanense*.

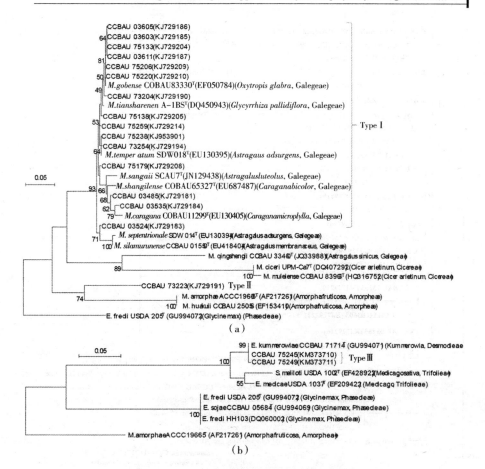

Fig. 2-5 ML phylogenetic trees of *nodC* gene sequences. Strains isolated from medical legumes in this study are shown in bold. Both the ML trees, subset (a) for *Mesorhizobium* and subset (b) for *Ensifer* species, were constructed with T92 + G model. Bootstrap values >50% are given at the branching points. Bars, 5% nucleotide substitution per site. Superscript (T) after strain number indicates the type strain of the species. Accession numbers of sequences for *nodC* genes extracted from GenBank are given in parentheses. Host plant and the tribe of the plant are included in parentheses. *E. fredii* or *M. amorphae* were used as outgroups respectively

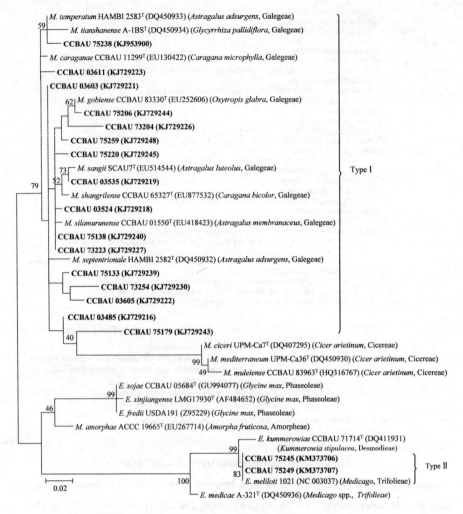

Fig. 2-6 ML phylogenetic tree of *nifH* gene sequences.
Isolates isolated from medicinal legumes in this study are shown in bold. T92 + G model was used to construct the phylogenetic tree. Bootstrap values ≥ 40% are given at the branching points. Bar, 2% nucleotide substitution per site. Superscript (T) after strain number indicates the type strain of the species. Accession numbers of sequences for *nifH* genes extracted from GenBank are given in the parentheses

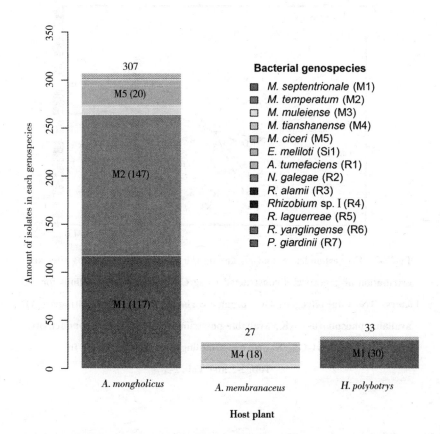

Fig. 2-7 Statistics of the isolates of bacterial genospecies isolated from root nodules of three medicinal legumes. This plot was created using the *boxplot* function integrated in the R language. The predominant genospecies (and the included isolates) from each legumes are specifically shown within the columns, as M1 (117), M2 (147) and M5 (20) in *A. mongholicus*; M4 (18) in *A. membranaceus* and M1 (30) in *H. polybotrys*. Total numbers of isolates from each medicinal legumes are shown above the bars

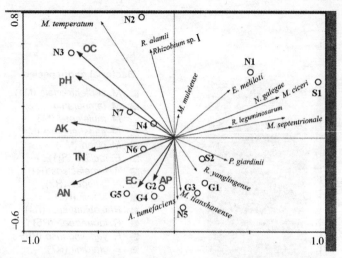

Fig. 2-8 Correspondence analysis among the soil factors, sampling sites and distribution of genospecies constructed using CANOCO. Abbreviations for soil factors: TN, total nitrogen; OC, organic carbon; AN, available nitrogen; AP, available phosphorus; AK, available potassium; EC, electrical conductivity. N1-N7, G1-G5, S1, S2 are sampling sites corresponding to Tab. 2-2 and Tab. 2-3

Discussion

Diversity of rhizobia associated with medicinal legumes

In this study, 367 bacterial isolates were obtained from root nodules of cultured *A. mongholicus*, *A. membranaceus* and *H. polybotrys*. Among these isolates, mesorhizobia possessing symbiotic genes could form effective nodules on *A. membranaceus* and *A. mongholicus*. 14 isolates belonging to *Agrobacterium*, *Rhizobium*, *Neorhizobium* and *Pararhizobium* were obtained mainly from nodules in sites where higher levels of fertilizer were applied (Tab. 2-3) and symbiotic genes could not be amplified from these isolates. Furthermore, these 14 isolates could not nodulate their original hosts. Two isolates belonging to *Ensifer*, CCBAU 75249 and CCBAU 75245, could not nodulate *Astragalus* spp. even though they had symbiotic genes. Therefore, these 14 isolates could be classified as non-symbiotic endophytes, as reported previously.

The identification of 347 isolates (accounting for 94.5%) as *Mesorhizobium* suggested that *A. membranaceus* and *A. mongholicus* were most associated with *Mesorhizobium* microsymbionts, partly in accordance with previous observations on other *Astragalus* species distributed in natural environments. The preference for symbiotic bacteria is also true for other legumes, such as chickpea predominantly associating with *Mesorhizobium* and alfalfa selecting *Ensifer* as their effective nitrogen fixation partners. However, the super predominance of *M. septentrionale* and *M. temperatum* as the most abundant groups (Tab. 2-4) detected in this study was different from a previous study that 40% of the rhizobia belonged to *Rhizobium* and *Ensifer* in natural conditions. In another study, only one isolate (CCBAU 71155) from *A. mongholicus* was identified as *Rhizobium* sp. These differences could be related to the distinct cultivation system including different soil and climate conditions, and other leguminous plants present in the vicinity.

Of the 33 bacterial isolates found in *H. polybotrys* nodules, 30 belonged to *Mesorhizobium*, while the remaining 3 isolates belonged to *Rhizobium*. This rhizobial composition was similar to those of the two *Astrgalus* species and the isolates from *H. polybotrys* could form effective nodules on *A. membranaceus*. Similar nodulation genes were found among the isolates from *Astragalus* spp. and *H. polybotrys*, supporting the cross nodulation of their rhizobia. For the 27 isolates from *A. membranaceus*, 23 belonged to *Mesorhizobium* with *M. tianshanense* (18 isolates) as the predominant rhizobia (Fig. 2-7). This phenomenon might be caused by the different host varieties and the differences in soil conditions where these medicinal legumes were grown.

The definition of only three *nodC* lineages and the predominance of *nodC* type I among the *Mesorhizobium* isolates, despite the identification of different genospecies [Fig. 2-5(a)], demonstrated that the two *Astragalus* species and *H. polybotrys hosted rhizobia with stringent selection for symbiotic genes, similar to chickpea but differed from soybean. The *nodC* type I gene was also found in *Mesorhizobium* strains isolated from other *Astragalus* species and related plants in the genera *Glycyrrhiza*, *Oxytropis* and *Caragana*, which all belonged to tribe Galegae, meaning that the phylogenetically related legumes might have similar preferences for nodulation genes and chromosomal background of their microsymbionts.

Previously, *M. muleiense* and *M. ciceri* were isolated from *Cicer arietinum*. In the present study, isolates belonging to the two *Mesorhizobium* species were also detected, with 96.1%-98.7% MLSA similarities to the corresponding type strains (Tab. 2-4). However, the isolates originating from *Astragalus* had symbiotic genes (*nodC* and *nifH*) highly similar to other *Mesorhizobium* genospecies isolated from the same hosts and differing from the two rhizobial species originating from *C. arietinum* (Fig. 2-5). The incongruence of the phylogenetic relationship between symbiotic and housekeeping genes (and 16S rRNA gene) (Figs. 2-3, 2-4, 2-5 and 2-6) implied lateral gene transfer among the *Mesorhizobium* species, similar to that reported in *Lotus*-nodulating and chickpea-nodulating *Mesorhizobium* species. This lateral gene transfer might be caused by the strong selective pressure on the symbiotic genes by the legumes and the adaptation to environmental conditions.

The predominant microsymbionts of the three kinds of medicinal legumes had some differences (Fig. 2-7). *A. mongholicus* was most commonly associated with *M. septentrionale* (M1), *M. temperatum* (M2) and *M. ciceri* (M5) as its partners, while *A. membranaceus* and *H. polybotrys* were associated with *M. tianshanense* (M4) and *M. septentrionale* (M1), respectively. Different rhizobial species should be considered when the inoculants are selected to inoculate these legumes in different fields.

Distribution of the rhizobia

The amounts and distribution of rhizobia associated with *Astragalus* spp. and *H. polybotrys* were unbalanced and differed from site to site in this study. As the predominant and most widespread genospecies, *M. temperatum* was distributed in nodules isolated from 13 sampling sites in Shanxi, Gansu and Ningxia. The other abundant genospecies was *M. septentrionale*, which collected from six sites in Shanxi, Gansu and Ningxia provinces (Tab. 2-1 and 2-4). The relatively wide distribution and high abundance of these species might imply their wider environmental adaptation and the higher competitive abilities than other species. Based on the wide adaptability of *M. temperatum* and *M. septentrionale* spp., candidate rhizobial inoculant strains should be screened from these two species for cultivated *Astragalus* and *Hedysarum* in China.

The nitrogen content in soil is an essential factor influencing the nitrogen fixation of rhizobium-legume symbionts. Soils in sampling site S1 had the lowest total nitrogen and available nitrogen contents, so nodulation of the legumes there was necessary, indicating the essential role of biological nitrogen fixation to the plants. The dominant rhizobial species in nodules in site S1 (*M. septentrionale* and *M. ciceri*) implied that these two rhizobial species might survive well in barren soils. The highest available nitrogen content was found in sample site G4 and only 5 rhizobial strains were isolated from nodules of this site, clear evidence of the inhibition of the formation of the root nodules with excessive fertilization in farmlands. In addition, other factors such as organic carbon content and pH influenced the distribution of *M. temperatum* in nodules of *A. mongholicus*. Soil EC had a substantial influence on the distribution of *M. tianshanense* associated with *A. membranaceus* in nodules of site N5.

In conclusion, cultivated *A. membranaceus*, *A. mongholicus* and *H. polybotrys* formed symbioses with *M. temperatum*, *M. septentrionale*, *M. muleiense*, *M. ciceri*, and *M. tianshanense*, which harbored similar *nod* gene types. *M. septen-*

trionale was predominant in nodules where the plants were grown the barren fields (low contents of available nitrogen and organic carbon), while *M. temperatum* was predominant nodules where the plants were grown in nitrogen-rich fields. This biogeographical pattern provided evidence that soil fertility was a possible determinant for the diversity of rhizobia associated with medicinal legumes during the domestication procedure.

Chapter III Evolutionarily conserved *nodE*, *nodO*, T1SS, and hydrogenase system in rhizobia of *Astragalus membranaceus* and *Caragana intermedia*

Introduction

Based on their tremendously important medicinal values and remarkable sand-fixing effects, *Astragalus* and *Caragana* species (belonging to tribe Galegeae and tribe Hydysareae, respectively) are widely cultivated in the northwest region of China. The diversity of rhizobia associated with these plants has been extensively studied. Previous researches revealed that both the cultivated and wild AC plants mainly nodulate with *Mesorhizobium* strains, while strains from other rhizobial genera occupied minor proportion in the nodules. However, an exception was found in a previous study in our laboratory that strain *R. yanglingense* CCBAU01603 has a more competitive nodulation ability than the representatives of *Mesorhizobium* species (*M. silamurunense* CCBAU01550, *M. silamurunense* CCBAU45272, *M. septentrionale* CCBAU01583, *M. amorphae* CCBAU01570, *M. caraganae* CCBAU01502) when *Caragana* plants were grown in sterile vermiculite. From these results, it could be seen that the nodulation specificity and competence of rhizobial strains on AC plants was different among different strains. However, the genetic characteristics that influence the nodulation phenotype of AC rhizobia is still unknown.

As reported in previous studies, the nodulation ability of rhizobia was determined by the symbiosis genes located on the symbiosis island/plasmid and type III secretion system. Studies on recognition and symbiotic specificity between legume hosts and rhizobia have made deep significance for understanding the interaction between the host plants and the microsymbionts. Currently, with the development of next-generation sequencing technology, the symbiotic characteristics of rhizobia are facing further studies of genomic analysis. Some whole genomes of strains within the genera *Sinorhizobium* and *Bradyrhizobium* that nodulate *Glycine max* or *Medicago* have been studied for their host specificity. However, the study of host specificity referring to the rhizobia that comprises two genera and associated with different hosts in a cross-nodulation group is still deficient. Considering the great morphology and phylogenetic differences between AC plants and AC rhizobia might be valuable candidates for further research to dig out the symbiotic specificity at the level of cross-nodulation group.

The major symbiotic rhizobia nodulating AC hosts were *Mesorhizobium* strains harboring conserved *nodC* genes differed from those in non-AC-nodulating strains, while the *nodC* of AC symbiotic strain *R. yanglingense* CCBAU01603 was not clustered closely to the AC-nodulating *Mesorhizobium* strains. Since *nodC* gene is an important determinant of chitin oligosaccharide chain of Nod factor that is a contributor to host specificity according to the study on *Sinorhizobium* (former *Rhizobium*) *meliloti*, the same preference of nodulation with AC plants by *R. yanglingense* CCBAU01603 and the *Mesorhizobium* strains might be determined by other symbiotic genes. Therefore, in order to better understand the mutual genomic characteristics of *Mesorhizobium* and *Rhizobium* strains that preferred AC plants at genus level, 7 rhizobia were whole-genome sequenced and analyzed for the nodulation genes, secretion systems and evolutionarily specific gene clusters.

Materials and Methods

Strains and their growth conditions

M. silamurunense CCBAU01550 and *M. silamurunense* CCBAU45272 originally isolated from *Astragalus membranaceus*, and *M. temperatum* CCBAU01399, *M. amorphae* CCBAU01570, *M. septentrionale* CCBAU01583, *M. caraganae* CCBAU01502 and *R. yanglingense* CCBAU01603 originally isolated from *Caragana intermedia* were used for the whole genome sequencing in this study. Nodulation tests performed in our laboratory further confirmed the nodulation ability of these strains with AC hosts. All these rhizobial strains grown on TY agar at 28 ℃ for 3-5 days, and cultivated in TY broth at 28 ℃ at 180 r · min^{-1} for 3 days.

DNA isolation

Cell pellets were collected from 1 mL bacterial cultures by centrifugation at 13 000 r · min^{-1} for 2 min. Then the cell pellets were resuspended in 1 mL sterile water for washing then centrifuged. The genomic DNA was extracted from the pellets using Wizard© Genomic DNA Purification Kit (Promega) according to the suggested procedure. Extracted DNA were temporarily maintained at -80 ℃ before being sent to the sequencing companies (BGI-Shenzhen, China).

Genome sequencing, assembly, and annotation

The genomes were sequenced on platform Illumina HiSeq 2000 (100 × sequencing depth) by BGI-Shenzhen China, and low-quality sequencing reads were filtered. SOAPdenovo v2.01 in Linux operating system was used to assemble the genomes and GapCloser v1.12 was used to close the gaps after assembly. Different Kmer values were tested from 17 to 91 (odd numbers) and the assembly results with the highest N50 values were reserved as the best assembly result. The whole genome sequences of the seven strains have been deposited in GenBank (Tab. 3-1). Glimmer v3.02 that was widely used in bacteria gene prediction was used to predict coding genes based upon interpolated Markov models. Sequences of predicted genes and proteins were extracted using Perl scripts. Then protein sequences were aligned against the non-redundant (nr) protein database of NCBI for annotation.

Tab. 3-1 The general genome information of the genomes used in this study.

Group	Strains	Host	GenBank biosample No.	Genome size/Mb	(G+C)/%	tRNA genes	Protein-coding sequences (CDSs)
AC rhizobia	R. yanglingense CCBAU01603*	C. intermedia	SAMN02584782	7.67	59.0	70	8 256
	M. silamurunense CCBAU01550*	A. membranaceus	SAMN02712003	7.07	62.7	49	7 496
	M. silamurunense CCBAU45272*	A. membranaceus	SAMN04278952	7.22	63.1	49	7 692
	M. septentrionale CCBAU01583*	C. intermedia	SAMN02584783	7.67	62.1	51	7 356
	M. amorphae CCBAU01570*	C. intermedia	SAMN02712002	7.37	61.2	48	7 141
	M. caraganae CCBAU01502*	C. intermedia	SAMN02585731	7.20	62.4	50	7 005
	M. temperatum CCBAU01399*	C. intermedia	SAMN02585732	7.42	62.4	62	8 604
Other hosts	M. metallidurans STM2683	Anthyllis vulneraria	SAMEA2272539	6.23	62.0	45	5 962
	M. ciceri bv. biserrulae WSM1271	Biserrula pelecinus	SAMN00713576	6.70	61.7	49	6 264
	M. ciceri ca181	Cicer arietinum	SAMN02470606	6.42	61.5	56	6 694
	M. loti MAFF303099	Lotus japonicus	SAMD00061086	7.60	60.6	54	7 281

Continued

Group	Strains	Host	GenBank biosample No.	Genome size/Mb	(G+C)/%	tRNA genes	Protein-coding sequences (CDSs)
	Mesorhizobium sp. WSM1293	Lotus sp.	SAMN02597200	6.94	61.8	52	6 706
	M. loti R88b	Lotus corniculatus	SAMN02597285	7.20	62.4	54	7 179
	M. opportunistum WSM2075	Biserrula pelecinus	SAMN00713576	6.88	62.9	53	6 508
	M. ciceri WSM4083	Bituminaria bituminosa	SAMN02584909	6.84	61.3	50	6 616
	R. etli CFN42	Phaseolus vulgaris	SAMN02603106	6.53	60.5	50	6 156
	Rhizobium sp. BR816	Leucaena leucosephala	SAMN02261311	6.95	60.4	55	6 752
	R. etli bv. mimosae Mim1	Mimosa affinis	SAMN02603105	7.20	60.4	51	6 792
	R. leguminosarum bv. trifolii WSM2304	Trifolium polymorphum	SAMN00000679	5.87	60.6	53	6 415

Continued

Group	Strains	Host	GenBank biosample No.	Genome size/Mb	(G+C)/%	tRNA genes	Protein-coding sequences (CDSs)
	Rhizobium sp. IRBG74	Sesbania cannabina	SAMEA3138824	5.46	58.7	54	5 478
	R. leguminosarum bv. phaseoli 4292	Phaseolus vulgaris	SAMN02261312	7.35	60.2	50	7 177
Out group	B. japonicum USDA6	Glycine max	SAMD00060992	9.21	63.7	57	8 409
	Cupriavidus taiwanensis LMG19424	Mimosa pudica	SAMEA3138280	6.48	65.03	63	5 654

Note: *, strains that sequenced in this study

Bioinformatics analysis

In this analysis, the whole genome sequences acquired in the present study and those of other 14 *Mesorhizobium* and *Rhizobium* strains downloaded from the NCBI database were included (see Tab. 3-1 for detail). Genome sequences of *B. japonicum* USDA6 and *Cupriavidus taiwanensis* LMG19424 were also included as out group in the phylogenetic analysis. PGAP (Pan-Genome Analysis Pipeline) was used to analyze the pan-genome and core-genome of tested strains. Specific genes of each group were extracted using Perl scripts according to the results of PGAP analysis.

All these strains, except *R. yanglingense* CCBAU01603 and the out group strains, were divided into 3 groups: AC-originating *Mesorhizobium* (ACiM, 6 strains), non-AC-originating *Mesorhizobium* (non-ACiM, 8 strains), and non-AC-originating *Rhizobium* (non-ACiR, 6 strains). To identify the phylogenetically conserved genes present in AC rhizobia (including ACiM and *R. yanglingense* CCBAU01603), 20 tested strains of three groups (ACiM, non-ACiM, non-ACiR) were aligned to the 8 256 proteins of *R. yanglingense* CCBAU01603 using phmmer program in HMMER3.0 package. Bit score of each gene was analyzed among the three groups using ANOVA in R software based on the FDR adjusted p-value with a cutoff value of 0.001, then the evolutionarily specific genes for AC rhizobia were identified according to the identity bit score and p-value. Alignment of homologous genes was performed with ClustalW2.0, and phylogenetic analysis was operated with PhyML3.0 and SplitsTree. The similarity of homologous genes was calculated using MEGA 6.06 software with p-distance method.

PCR universal verification for *nodO* genes

To verify whether *nodO* gene was universal in other common AC rhizobia, PCR verification was performed using primers nodO-131F (5'-GCGAGGGCAGT-

GACCAA-3′) and nodO-554R (5′-GCCGCACCGCTGTAGAA-3′). This pair of primers was designed according to the *nodO* sequence of strain CCBAU45272 using the Primer Premier 5.0 software. The PCR reaction system was 50 μL, including Taq PCR Mix (23 μL), 131F (1 μL), 554R (1 μL), template DNA (1 μL) and ddH$_2$O (24 μL). The PCR protocol was 95 ℃ (5 min), 30 cycles [94 ℃ (50 s); 56 ℃ (50 s); 71 ℃ (50 s)], and 72 ℃ (5 min).

Construction of *nodO* mutant of CCBAU01603

Upstream fragment (727 bp) of the target gene was amplified using primers *nodO* up F: EcoR I -5′-CGGAATTC-GGCAAACATTTACCGACCGACTA-3′ and *nodO* up R: Kpn I -5′-GGGGTACC-CTTTACCATCGCAAACACTCCT-3′. Downstream fragment (812 bp) of the target gene was amplified using primers: *nodO* down F: snaB I -5′-GACTTACGTA-AGTTCGTTCACCTGAGCGG-3′ and *nodO* down R: sac I -5′-GACGAGCTC-CGCGTTGAAAGCGGAAG-3′. The PCR system and protocol was the same as mentioned above of *nodO* universal fragment. After amplification, the empty vector plasmid pRL1063a and upstream fragment were digested with double endonuclease restriction (*Eco*R I and *Kpn* I) for 4 h, respectively. Double digested plasmids and upstream fragment were linked by T4 DNA ligase at 16 ℃ for 8 h. The upstream fragment containing plasmid was transformed into strain *E. coli* DH5α using heat shock method, then the integrated DH5α strain was incubated on LB plating medium for 24 h. Several single colony isolates were inoculated in TY broth in 37 ℃ at 180 r · min^{-1} for 8 h, and then the strains were delivered to sequencing company to screen the isolate that without nucleotide mutations. The same method was used to link the downstream fragment to plasmid pRL1063a. Then the donor bacteria that carried vector containing upstream fragment and downstream fragment has been constructed.

Homologous pair exchange method was used to construct *nodO* deletion mutant of strain CCBAU01603. Donor strain, accessory strain and receptor strain were fully mixed according to ratio of 3:2:10, and the mixture was spread on TY plates to incubate at 28 ℃ for 24-48 h. Then the cultures were scraped into 1 mL of physiological saline and diluted to 10^{-6}-10^{-8}. Aliquot of 0.1 mL of the last three dilutions was spread on plates with TY medium. After incubated for 48 h at 28 ℃, single colony isolates were picked out and inoculated on TY medium supplied with 5 μg·mL^{-1} tetracycline for further incubation at 28 ℃ for another 48 h. Single colony isolates lacking the resistance to tetracycline were obtained. Then the *nodO* deletion mutants were verified by absence of *nodO* amplification with the methods mentioned above.

Inoculation test and the observation of bacteroids

Wild type and *nodO* deletion mutant of strain CCBAU01603 were cultured in 5 mL of TY broth at 28 ℃ for two days and were inoculated on hosts *A. membranaceus*, *C. intermedia* and the promiscuous legume *Sophora flavescens* separately using the protocol described previously. The plants were cultivated in vermiculite for 40 days in greenhouse with a cycle of 16 h light at 25 ℃ and 8 h dark at 22 ℃. Five replicates were repeated. Then the nodules of wild type and *nodO* mutant were obtained and fixed in 2.5% glutaraldehyde solution immediately. The fixed nodules were delivered to the Electron Microscope Lab to do further technological process, then the bacteroids were observed at 2 500 × and 10 000 × magnification using transmission electron microscope in China Agriculture University.

Results

The general features of the genomes used in this study

As results (Tab. 3-1), the genome sizes of AC-originating strains were about 7.07-7.67 Mb, which were generally larger than those of non-AC-originating strains that presented genome sizes 6.23-7.60 Mb (t-test p-value = 0.011 6).

Likewise, the CDS numbers present a good correlation with the genome size ($R^2 = 0.979$), and the number of predicted CDSs of AC-originating strains were obviously higher than that of the non-AC-originating strains. Besides, the DNA G + C contents of ACiM strains (61.2%-63.1%) were significantly higher than those of the non-AC-originating strains (60.2%-62.0%; t-test, p-value = 0.02), excepts that DNA G + C content of *R. yanglingense* CCBAU01603 was 59.0 %.

Our analysis revealed that the 21 researched rhizobia (14 *Mesorhizobium* and 7 *Rhizobium*) strains contained a pan-genome of 29 274 putative protein-coding genes. Moreover, the 21 genomes shared a core genome of 823 genes, accounting for 9.6%-15.0% of the whole genomes, representing a set of conserved orthologous genes for both *Mesorhizobium* and *Rhizobium* genera. Out of these 823 core genes, 362 single copy core genes that represents the reliable genetic relationship of 21 rhizobia were used to perform a maximum likelihood phylogenetic analysis, and results showed that rhizobia of *Mesorhizobium* and *Rhizobium* evolved divergently as two different genera (Fig. 3-1). It's true that *R. yanglingense* CCBAU01603 clustered with other *Rhizobium* strains, and all the *Mesorhizobium* strains were phylogenetically clustered in a conserved branch.

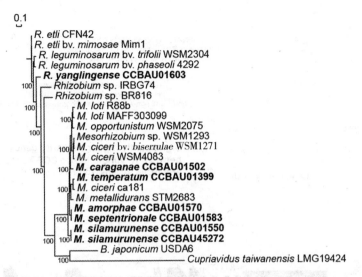

Fig. 3-1 Phylogenetic relationships of AC-isolated and non-AC-isolated strains based on 362 single copy core genes. Maximum likelihood method was used to construct the tree using PhyML3.0. Total 362 single copy core genes were exacted according to the PGAP results and connected after alignment using ClustalW2.0. Bold represents the AC rhizobia that sequenced in this research

Conserved Genes in AC-originating strains

Using the program phmmer in HMMER3.0 that based on the alignment of protein conserved domain, a set of 130 genes that conserved for ACiM strains and *R. yanglingense* CCBAU01603 were screened out, which had significantly higher similarities between ACiM and strain CCBAU01603 than non-ACiM and non-ACiR groups, and were presumed as homologous genes that have closer evolutional relationships between ACiM group strains and *R. yanglingense* CCBAU01603, and may be responsible for nodulation specificity preference [based on FDR value < 0.001, Fig. 3-2(a)]. Interestingly, these 130 genes derived from 5 scaffolds, out of which, a scaffold contains the nodulation island including evolutionarily specific *nodZ*, *nodE*, *nodO* and T1SS for all these seven AC-originating strains [Fig. 3-2(b)].

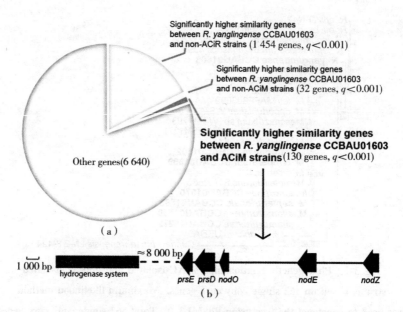

Fig. 3-2 Analysis of evolutionarily closer genes in strains of these three groups (ACiM, non-ACiM, and non-ACiR) in comparison to *R. yanglingense* CCBAU01603. (a) Evolutionarily closer genes that shared higher genetic similarities in comparison with *R. yanglingense* CCBAU01603. Data in parentheses represent numbers of shared evolutionarily closer genes. Bold was evolutionarily closer genes for AC rhizobia including two genera (*Mesorhizobium* and *Rhizobium*). q-value threshold was 0.001. (b) Arrangement of evolutionarily closer genes between ACiM and *R. yanglingense* CCBAU01603, including evolutionarily specific *nodE*, *nodO*, T1SS and hydrogenase system. Dash line was a region of about 8 kb

Conserved nodulation genes in AC-originating strains

Since nodulation genes are the key determinants in host recognition and signal transduction in rhizobia, some nodulation genes were comparatively studied among the AC microsymbionts. *R. yanglingense* CCBAU01603 and other 6 ACiM strains shared the highly evolutionarily conserved *nodE*, even multiple copies of *nodE* were detected in *R. yanglingense* CCBAU01603 [Fig. 3-3(a)]. Besides, *nodC* genes of ACiM strains were phylogenetically distinctive from those of other non AC rhizobia, and also divergent from that of *R. yanglingense* CCBAU01603 [Fig. 3-3(b)]. Even though, the *nodC* of *R. yanglingense* CCBAU01603 shared relatively higher similarity with ACiM (p-distance of NodC: 0.181-0.194) compared to non-ACiM and non-ACiR groups (p-distance of NodC: 0.200-0.308). In addition, an uncommon gene *nodZ* was identified in all the 7 AC-originating strains, but it evolved divergently in *R. yanglingense* CCBAU01603 compared with those in the 6 AC-originating *Mesorhizobium* strains [Fig. 3-3(c)].

Conserved hydrogenase systems and widely spread *nodO* genes in AC-originating strains

According to the results of phmmer alignment analysis, a system that responsible for hydrogenase biosynthesis (*hup*) was detected in the 7 AC-originating strains [Fig. 3-4(a)], which was supposed to be positively related to the nitrogen fixation efficiency in some rhizobial strains. In the other 14 reference strains, hydrogenase systems only have been detected in *Rhizobium* sp. BR816, *R. etli* bv. *mimosae* Mim1 and *M. opportunistum* WSM2075. Interestingly, the hydrogenase systems were found to be evolved from two ancestries in these 7 AC-originating strains: *M. caraganae* CCBAU01502 and *R. yanglingense* CCBAU01603 shared a mutual origin (p-distance: 0.010), which presented closer genetic distance from the other non-AC-originating strains (p-distance: 0.050-0.101) than the other five ACiM strains [p-distance: 0.193-0.200; Fig. 3-4(b) and (c)].

Moreover, specifically conserved *nodO* as a host range expanded gene was also detected in all the seven AC-originating strains. Since *nodO* also presented in the broad host strains *Rhizobium* sp. BR816 and *R. etli* bv. *mimosae* Mim1, it may contribute to the broad host range phenotype in rhizobial strains. Therefore, more *nodO* sequences were downloaded from GenBank database to analyze their phylogenetic relationships with those AC-originating strains. As shown in Fig. 3-5, the *nodO* genes of the 7 AC-originating strains were obviously divergent from those of the other rhizobia, and *R. yanglingense* CCBAU01603 harbored a *nodO* gene that shared extremely high similarity (95.6% - 98.8% nucleotide similarity) with the 6 ACiM strains. In addition, the *nodO* genes of AC-originating rhizobia shared only 46.7% - 75.8% similarities with those in other rhizobia. Moreover, *nodO* was exactly amplified out in all 14 representative strains (with nodulation ability) of 367 isolates from *Astragalus* grown in 3 provinces of China using universal primers of *nodO* (Tab. 3-2).

The T3SS and conserved T1SS of AC-originating rhizobia

For the strains studied in this research, 4 to 8 pairs of *prsD/prsE* genes (T1SS structural genes) were identified in these 7 AC-originating strains, but only 2 to 4 pairs were discovered in the other strains. Out of which, a pair of evolutionarily conserved and private *prsD/prsE* genes were detected right at the downstream of *nodE* and *nodO* genes and the upstream of *hup* biosynthesis system gene cluster. And this pair of *prsD/prsE* genes in the 7 AC-originating strains evolved distinctively from non-ACiM and non-ACiR rhizobia, which indicated a homologous and consistent relationship with its adjacent *nodE* and *nodO* genes [Fig. 3-6 (a) and (b)].

However, typical T3SS was identified for strains *M. silamurunense* CCBAU01550, *M. silamurunense* CCBAU45272 and *M. temperatum* CCBAU01399, and in the phylogenetic trees of *rhcJ* and *rhcS* genes (structural genes of T3SS), these three strains were grouped together with type *rhc*-II T3SS of *Rhizobium* sp. BR816 and *Sinorhizobium fredii* NGR234. While the remaining AC-rhizobia harbored atypical T3SSs (more similar to flagellum biosynthesis system) and clustered with other *Mesorhizobium* or *Rhizobium*, according to their genus affiliation [Fig. 3-6(c) and (d)].

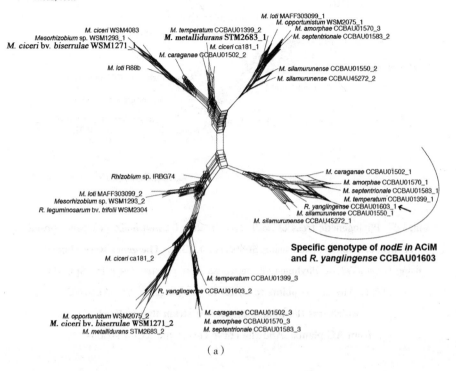

(a)

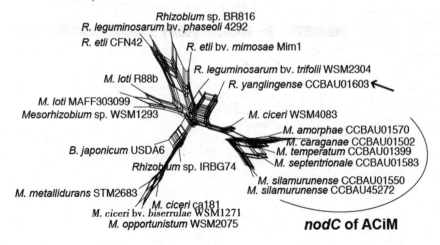

(b)

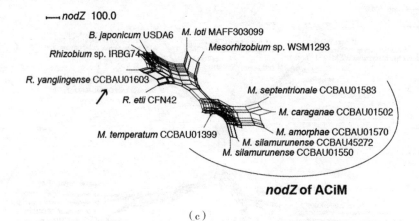

(c)

Fig. 3-3　Phylogenetic trees of *nodE* (a), *nodC* (b) and *nodZ* (c) genes based on network construction using SplitsTree 4.13.1. The genes were aligned using ClustalW2.0. Phylogenetic networks were constructed using SplitsTree 4.13.1. The arrow points to strain *R. yanglingense* CCBAU01603, which was the unique *Rhizobium* strain that isolated from AC plants. And the curve covers the ACiM strains

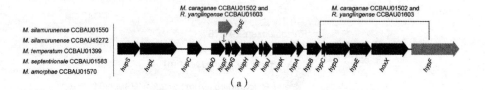

(a)

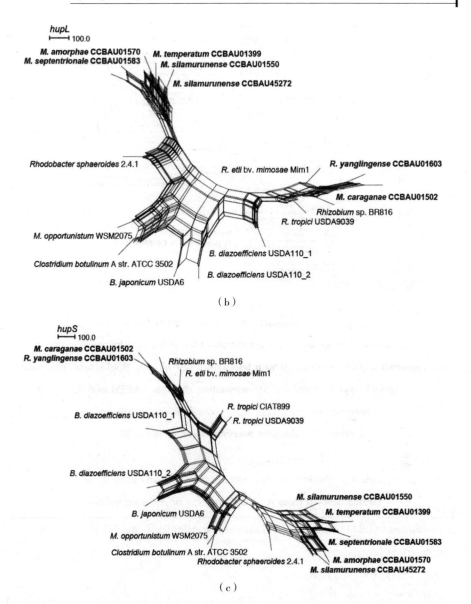

Fig. 3-4 Arrangement of hydrogenase system gene clusters (a) and phylogenetic relationships of *hupL* (b) and *hupS* (c) of the AC-isolated (Bold) and other strains (Regular). (a) These genes arranged from *hupS* to *hypF* without *hupE* for strains CCBAU01550, CCBAU45272, CCBAU01399, CCBAU01583, and CCBAU01570. A *hupE* gene was inserted between *hupD* and *hupF*, and *hypF* translocated for strains CCBAU01502 and CCBAU01603. (b), (c) The genes were aligned using ClustalW2.0. Phylogenetic networks were built using SplitsTree 4.13.1

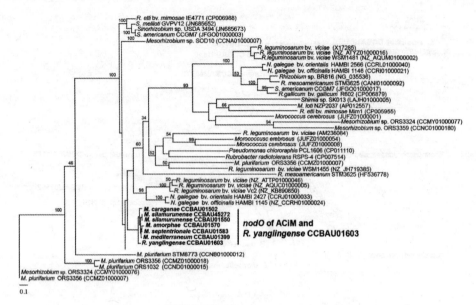

Fig. 3-5 Phylogenetic tree of *nodO* genes based on ML method. The genes were aligned using ClustalW2.0. Phylogenetic tree was constructed using PhyML3.0 with bootstrap value of 100. Bold strains represent specific type of *nodO* for AC-originating rhizobia (ACiM and *R. yanglingense* CCBAU01603). Numbers in parentheses represent the gene accession number in NCBI

Tab. 3-2 Nodulation and amplification of *nodO* gene from strains isolated from *A. membranaceus* or *A. mongholicus*

Strains (CCBAU No.)	Nodulation capability	Amplification of *nodO*
75179, 03611, 03605, 73254, 75138, 75133, 03603, 75206, 75220, 03535, 03524, 75238, 75213, 73204	+	+
75228, 73231, 75219, 75237, 03532, 03529, 73239, 75233, 75168, 75186, 75229, 73262, 75204	−	−

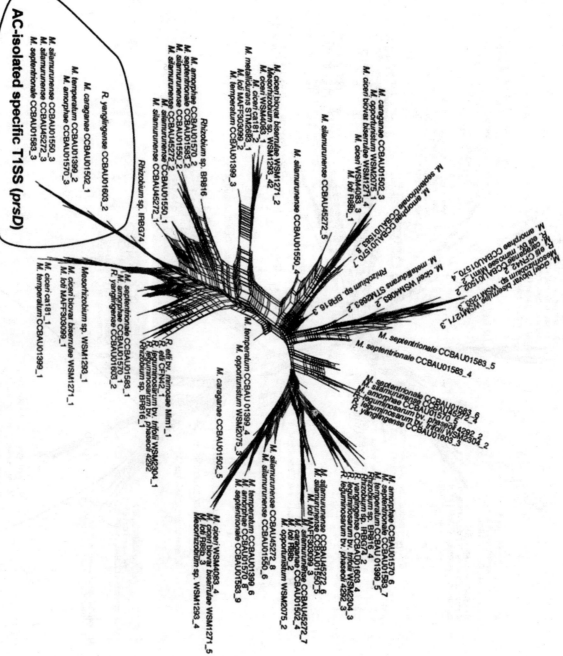

(a)

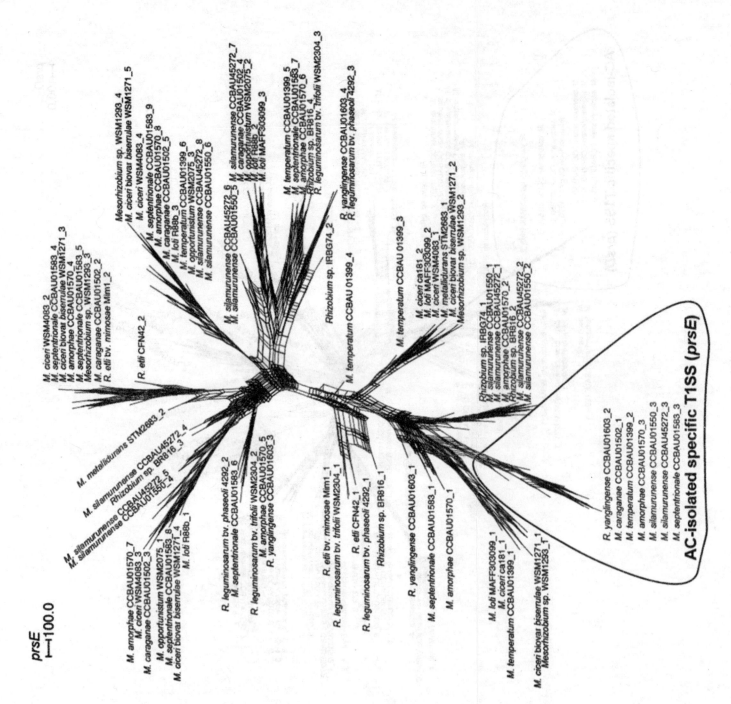

(b)

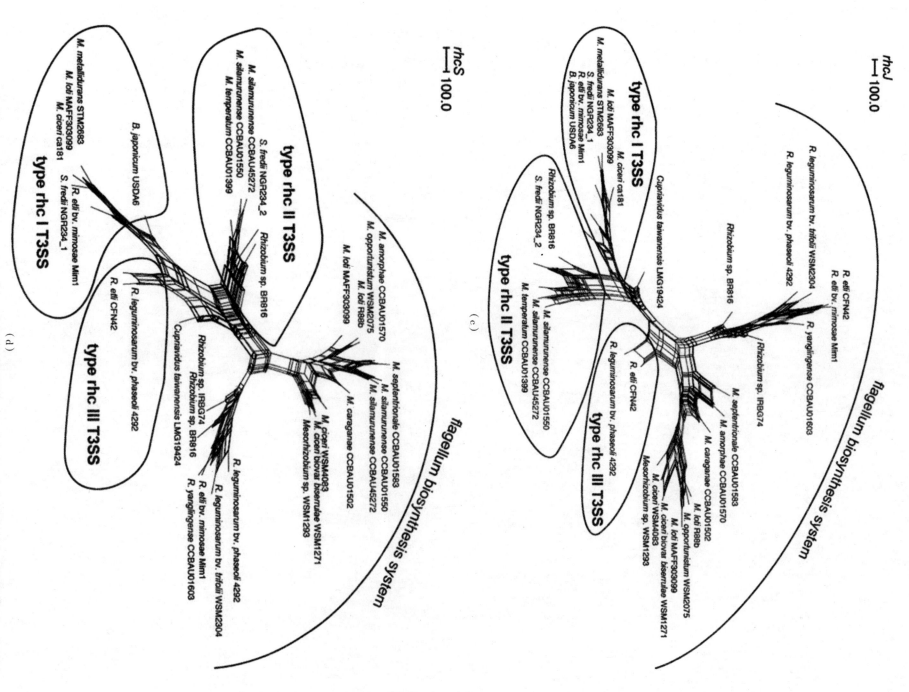

Fig. 3-6 Phylogenetic trees of T1SS (*prsD* and *prsE*) and reserved T3SS genes (*rhcJ* and *rhcS*) based on SplitsTree network. The genes were detected using phmmer in HMMER3 with E-value and coverage were 1e-5 and 50%, respectively. The genes were aligned using ClustalW2.0. ML phylogenetic networks were constructed using SplitsTree 4.13.1. (a), *prsD* (T1SS); (b), *prsE* (T1SS); (c), *rhcJ* (T3SS); (d), *rhcS* (T3SS)

The nodulation test of CCBAU01603 wild type strain and its *nodO* mutant

A *nodO* deletion mutant of CCBAU01603 was constructed to verify the impact of *nodO* on symbiosis. The wild type of CCBAU01603 could form effective nodules on *A. membranaceus*, *C. intermedia* and *S. flavescens*. But, the *nodO* deletion mutant formed ineffective small nodules on *A. membranaceus*, which present white or black nodule sections [Fig. 3-7(a)]. However, the nodules formed by mutant of CCBAU01603 on *C. intermedia* and *S. flavescens* were normal in morphology [Fig. 3-7(b) and (c)]. The observation of electron microscopy on section of nodules showed that the symbiosome membrane in the nodules formed by the mutant on *A. membranaceus* and *C. intermedia* were both severely disrupted, the plant cells presented plasmolysis phenomenon, and the bacteroids showed abnormal morphology [Fig. 3-7(d) and (e)]. Nevertheless, the normal symbiosome membrane and morphology of bacteroids in nodules [Fig. 3-7(c) and (f)] induced by the mutant on *S. flavescens* may be ascribed to the extremely promiscuous property of this legume.

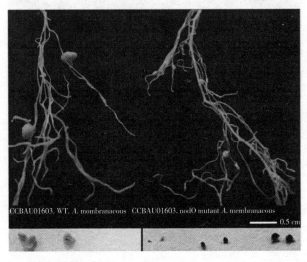

(a)

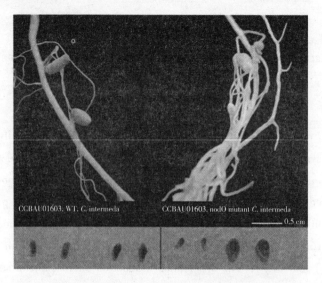

(b)

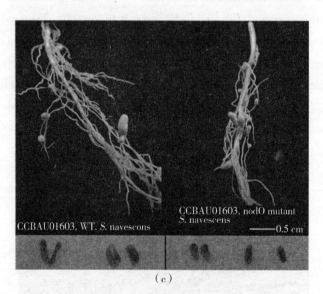

(c)

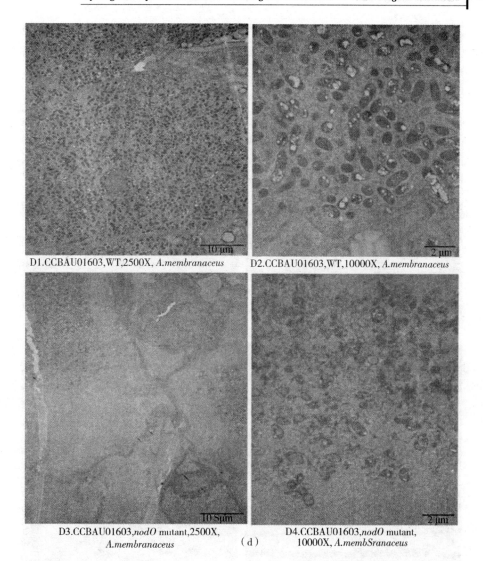

D1.CCBAU01603,WT,2500X, *A.membranaceus*
D2.CCBAU01603,WT,10000X, *A.membranaceus*
D3.CCBAU01603,*nodO* mutant,2500X, *A.membranaceus*
(d)
D4.CCBAU01603,*nodO* mutant, 10000X, *A.membSranaceus*

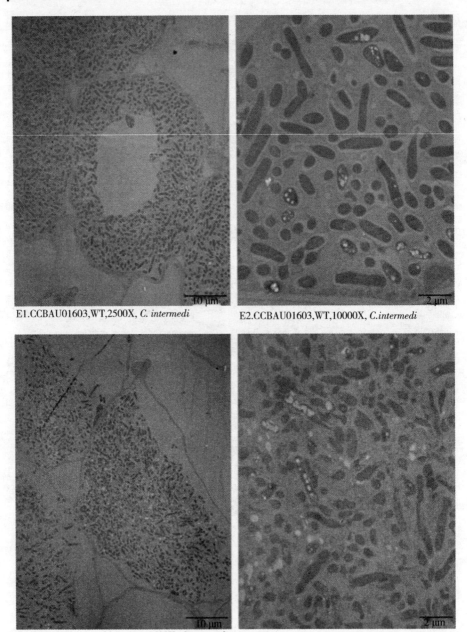

E1.CCBAU01603,WT,2500X, *C. intermedi*   E2.CCBAU01603,WT,10000X, *C.intermedi*

E3.CCBAU01603,*nodO* mutant, 2500X, *C.intermedi*   E4.CCBAU01603,*nodO* mutant, 10000X, *C.intermedi*

(e)

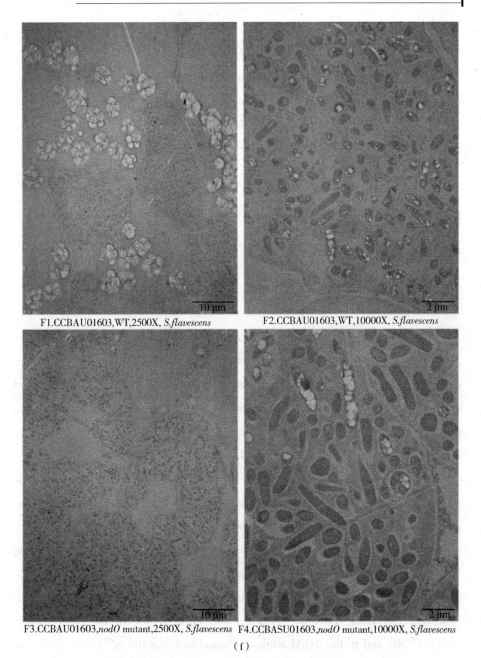

(f)

Fig. 3-7 Nodulation phonotype and bacteroids morphology of *nodO* mutant of *R. yanglingense* CCBAU01603 under transmission electron microscope (TEM) (a)-(c): Nodulation phonotype on *A. membranaceus* (a), *C. intermedia* (b), and *S. flavescens* (c), respectively.

(d)-(f): Bacteroids morphology of strains inoculated to *A. membranaceus* (d), *C. intermedia* (e), and *S. flavescens* (f), respectively. d1, d2, e1, e2, f1, f2: CCBAU01603 wild type. d3, d4, e3, e4, f3, f4: CCBAU01603 *nodO* mutant. d1, d3, e1, e3, f1, f3: 2 500 ×. d2, d4, e2, e4, f2, f4: 10 000 ×.

WT: wild type.

Discussion

Specific genotype of *nodE* and *nodZ* for AC rhizobia

The *nodE* gene has been proved as a major determinant of the host specificity in *S. meliloti* and *R. leguminosarum*, which is responsible for the synthesis of the acyl moiety of Nod factors. In the present study the detection of highly conserved *nodE* in both the ACiM and *R. yanglingense* CCBAU01603 implying that this gene might be also the determinant for their host specificity, and that the AC plants may recognize Nod factor with same fatty acyl groups.

According to the previous research, the nodulation gene *nodZ*, which encodes a lipochitinoligosaccharide (nodulation factors) fucosyltransferase, was a host-specific gene in *B. japonicum*, *Sinorhizobium* sp. NGR234, and *M. loti*. In the present research, the identification of highly conserved *nodZ* in the six AC-originating *Mesorhizobium* strains and their difference from other *Mesorhizobium* strains evidenced that this gene may contribute to the host specificity in the AC rhizobia. However, the great difference between the *nodZ* genes in *R. yanglingense* CCBAU01603 and in the ACiM strains demonstrated that the AC plants might recognize *nod* factors with various molecular structures, or both *nodZ* gene types could generate the same nod factors, for which further study to estimate the nod factor patterns for both the ACiM and *R. yanglingense* CCBAU01603 is needed.

Highly conserved hydrogenase systems of AC-originating strains

Hydrogen uptake that accompanying the nitrogen reduction could increase the nitrogen fixation efficiency by recycling the hydrogen for nitrogen reduction. The hydrogenase structural and most assessor genes are only expressed under the control of the nitrogen fixation regulatory protein NifA. Thus, the hydrogenase biosynthesis was relied upon nodulation gene expression and nitrogen fixation. Hence, there may be some relationships between the hydrogenase system and host specificity preference in the AC rhizobia. Previously, hydrogen oxidation ability was discovered in *Bradyrhizobium* sp. (*Lupines*), *Bradyrhizobium* sp. (*Vigna*), a few strains in *R. leguminosarum* bv. *viciae* and *R. tropici*, but was never reported in *Mesorhizobium*. To our best knowledge, hydrogenase systems were common in *Bradyrhizobium*, but rarely identified in *Mesorhizobium*, *Rhizobium* and *Sinorhizobium*. Hence, the prevalent conservation of hydrogenase biosynthesis system in the AC rhizobia (including *Mesorhizobium* and *Rhizobium*) may be an environmental or host adaptation strategy. As reported in previous research, pivotal and conserved hydrogenase genes *hupS* and *hupL* showed great divergence among the *Rhizobium* and *Bradyrhizobium* strains, implying that their hydrogenase systems may be obtained from different sources or have evolved independently. But, in our study, although the hydrogenase systems prevalently exist in ACiM strains evolved distinctively from *Bradyrhizobium* and *Rhizobium*, which representing a novel lineage of hydrogenase system, but the hydrogenase system of *M. caraganae* CCBAU01502 evolved divergently from other *Mesorhizobium* strains, and keep high homology with *R. yanglingense* CCBAU01603 and other *Rhizobium* strains. These results indicates that the hydrogenase system does not stably coevolve with house-keeping core genes, and could horizontally transfer among different genera, and the existence of hydrogenase in the AC rhizobia may not related to their host specificity, but to the nitrogen fixation efficiency.

Highly conserved *nodO* genes in AC rhizobia

The *nodO* gene is partially homologous to the hemolysin gene in *E. coli* and it located at the downstream of *nodDEFDABCIJZ* gene cluster. Similar to hemolysin, NodO protein could be secreted on the growth medium independent on the flavonoids synthesis by the type I secretion system, but some study showed that NodO was only produced when bacteria grew in the presence of a flavonoid *nod* gene inducer. Anyway, *nodO* was defined as a mutual feature that shared among ACiM and strain *R. yanglingense* CCBAU01603, and also defined in 14 representative strains of 367 isolates from *Astragalus*, being a common feature for AC rhizobia. Up to now, *nodO* has only been reported in *R. leguminosarum* bv. *viciae* and is proved responsible for host range of vetch, *Leucaena leucocephala*, *Phaseolus vulgaris* and *Trifolium repens*. Moreover, the heterologous expression of *nodO* could extend the host range of rhizobia and enable *R. leguminosarum* bv. *trifolii* to nodulate vetch. Also, *nodO* gene has been identified as nodulation and host-specific recognition factor of rhizobia associated with pea and *Vicia* spp. in *R. leguminosarum* bv. *viciae*. Indeed, the *nodO* gene in *R. leguminosarum* bv. *viciae* encodes a hemolysin homologous Ca^{2+}-binding protein without N-terminal cleavage, and NodO protein could trigger cation-selective channels that allow K^+ and Na^+ across the cell membranes. Therefore, NodO may facilitate the uptake of *nod* factors or function synergistically with depolarization or complete the deficiencies in Nod factor signaling. Hence, *nodO* is presumed to participate in recognition and signal transduction for rhizobia to nodulate AC plants. Interestingly, this estimation related two previous reports together, *N. galegae* that could nodulate *Astragalus* plants also contained a *nodO* gene. Although a unique lineage of *nodO* was detected in all the AC-nodulating strains but failed in the other strains in our study, this gene might not be directly related to the host specificity according to the nodulation tests with the *R. yanglingense* CCBAU01603 *nodO* mutant. Nevertheless, the deletion of *nodO* gene affected the development of bacteroids in nodules of the AC plants, indicating that *nodO* has an impact on the nodule formation process, but not the recognition process.

Highly conserved T1SS in AC rhizobia

Previous studies have proved that various protein secretion systems were used in rhizobia to transport effector proteins involved in nodulation process, and *nodO* was secreted via type Ⅰ secretion system (T1SS). T1SS could secrete proteins from the bacterial cytoplasm to extracellular environment, and was responsible for the secretion of various toxins, lipases and proteases in many Gram-negative bacteria. T1SS has been verified to be responsible in secretion of NodO and other homologous proteins of hemolysin without N-terminal cleavage. In our present research, 4 to 8 pairs of T1SS structural genes *prsD/prsE* were detected in 7 AC rhizobia. Out of which, we found that a pair of *prsD/prsE* that maintained high conservation among AC rhizobia (96.4%-100% similarity) located rightly besides the conserved *nodO* gene. This pair of *prsD/prsE* genes in AC rhizobia kept low similarity (55.0%-70.8%) with other non AC rhizobia, which indicated that the AC rhizobia has a specific T1SS system for nodO. The adjacent location of *prsD/prsE* genes to *nodE*, *nodO* and hydrogenase system in the AC rhizobia implied that *nodO* gene in AC-isolated strains is coevolved with its adjacent T1SS, and NodO protein may be secreted specifically through the nearby *prsD/prsE* secretion system at downstream.

It is well known that T3SS plays a significant role in biochemical cross-talk between bacteria (including animal- and plant-pathogen) and eukaryotic hosts. However, it had been supposed to be dispensable for nodulation of rhizobia, and it may positively or negatively affect the nodulation process. With the development of sequencing technology, rhizobial nodulation without T3SS was found increasing. In rhizobia, the rhc gene clusters that encode T3SS-related components could be divided into three subtypes: rhc-Ⅰ in *S. fredii* NGR234 and *B. japonicum* USDA110; rhc-Ⅱ in *S. fredii* NGR234; and rhc-Ⅲ in *R. etli* CFN42. Since rhc-Ⅱ type T3SS was identified in 3/7 AC-nodulating rhizobial strains and non-AC-nodulating strain *S. fredii* NGR234, while atypical T3SS was detected in the other four AC-nodulating rhizobial strains, it might be estimated that the rhc-II type T3SS may not be involved in the specific nodulation on AC plants. Moreover, the finding of atypical T3SS lineages noted as flagellar synthesis genes (annotated using the Non-Redundant Protein Database) in the 4 AC rhizobia demonstrated the possibility that the flagellar synthesis serves a function as a novel type of T3SS for AC rhizobia.

Conclusion

Comparative genomics revealed that a large conserved fragment of functional gene clusters including evolutionarily conserved *nodE*, *nodO*, T1SS, and hydrogenase system were detected in the *Astragalus/Caragana*-associating *Mesorhizobium* and *Rhizobium* strains, and the *nodO* gene was found prevalently exists in common AC rhizobia. These genes were deduced to participate host recognition of AC rhizobia. More profoundly, the existence of particular T1SS is inferred as a factor to regulate the host specificity in these strains. These results revealed that there exist several extremely relevant genomic influencing factors for the preference between host and their rhizobia, and provided substantial materials for further research on the specificity of symbiosis between rhizobia and legumes at the cross-nodulation level.

Chapter IV Genetic divergence among *Bradyrhizobium* strains nodulating wild and cultivated *Kummerowia* spp. in China

Introduction

Kummerowia stipulacea (Korean clover) and *K. striata* (Japanese clover) are legume grasses widely distributed in roadside, hillside, lakeside and other wild fields (WFs) habitats in China and other countries. Because of its high oestrogenic activity and production of flavonoids used for menopausal dysfunction treatment, anti-HIV and antitumor medicine, herbal *K. striata* has been deliberately cultivated. Futhermore, *Kummerowia* is introduced as a flowering grass in university campuses in urban areas (UAs). Symbiotic nitrogen fixation of *Kummerowia* with rhizobia plays an important role in the healthy growth of the plant without fertiliser application. Rhizobial strains of *Kummerowia* have been recorded previously, of which slow-growing strains such as *Bradyrhizobium* spp. are the predominant microsymbionts in China. Some fast-growing strains grown in the Loess Plateau have been identified as *Sinorhizobium kummerowiae* and in campus lawns as *Rhizobium cauense*. Furthermore, 13 reference strains (*B. diazoefficiens*, *B. canariense*, *B. japonicum*, *B. betae*, *B. huanghuaihaiense*, *B. daqingense*, *B. liaoningense*, *B. yuanmingense*, *B. iriomotense*, *B. elkanii*, *B. pachyrhizi*, *B. jicamae* and *B. lablabi*) have been employed to identify the phylogenetic relationship of these rhizobial strains isolated from nodules on the roots of *Kummerowia* distributed in WFs and UAs.

So far, some novel species of *Bradyrhizobium* have been classified. The biodiversity of the rhizobial strains of *Kummerowia* associated with biogeographic patterns has also been recently studied, but the genetic communication underlying their long-term evolutionary histories has not yet been elucidated. An important preparation is to distinguish the rhizobial species and even the subdivisions of each species systematically based on taxonomic markers, especially when multi-genes are concatenated. Natural selection and genetic drift based on mutation, gene flow, rearrangement and recombination of the genome of these rhizobia in the microevolution process could then contribute to the understanding of the mechanisms of biodiversity generation and preservation. Rhizobial strains associated with *Kummerowia* isolated both from WFs and UAs in this study might be the best material to explore this microevolutionary mechanism. Recent studies have shown that the three taxonomic markers (*SMc00019*, *truA* and *thrA*) are useful for constructing phylogenetic trees with a gap of 2% for the similarity percentage of intraspecies/interspecies average nucleotide identity (ANI). This will support the predetermined species, thereby corresponding to a taxonomy system based on the rhizobial genomes. Three ecoregion-related genes (*atpD*, *glnII* and *recA*), important housekeeping core genes, however, have not been compared with the taxonomic markers (*SMc00019*, *truA* and *thrA*) in previous studies.

In this study, evolutionary analyses of *Bradyrhizobium* strains nodulating *Kummerowia* collected from WFs and UAs in China were performed by studying three genes (*SMc00019*, *truA* and *thrA*) as taxonomic markers, ecoregion-related genes (*atpD*, *glnII* and *recA*) as environmental conditions adapters and two critical symbiosis-related genes (*nodC* and *nifH*) as host legume selectors. Biodiversity, genetic divergence, gene flow, recombination and evolutionary lineages were compared among the *Kummerowia* bradyrhizobia isolated from WFs and UAs.

Materials and Methods

Rhizobial strains

A total of 283 rhizobial isolates from root nodules of *Kummerowia* were collected from 28 sites in southern (99 isolates, 6 provinces) and northern (184 isolates, 2 provinces). Geographically, these sites can be divided into two segregated ecoregions corresponding to WFs and UAs environmental conditions, with 130 isolates and 153 isolates, respectively. These strains were previously preserved in the Culture Collection of Beijing Agricultural University (CCBAU) in tryptone-yeast (TY) extract medium with 20% glycerol. Thirteen reference strains were used to determine the phylogenetic positions of the *Kummerowia* bradyrhizobia: *B. betae* LMG 21987^T, *B. canariense* BTA-1^T, *B. daqingense* CCBAU 15774^T, *B. elkanii* USDA 76^T, *B. huanghuaihaiense* CCBAU 23303^T, *B. iriomotense* LMG 24129^T, *B. japonicum* USDA 6^T, *B. diazoefficiens* USDA 110^T, *B. jicamae* PAC68^T, *B. lablabi* CCBAU 23086^T, *B. liaoningense* USDA 3622^T, *B. pachyrhizi* PAC48^T and *B. yuanmingense* CCBAU 10071^T. All the strains were grown in YMA medium at 28 °C.

Gene amplification and sequencing

The total template DNA of these strains was extracted using the method described by Terefework et al. The primer sequences and PCR protocols used for the amplification of three taxonomic markers (*SMc*00019, *truA* and *thrA*), ecoregion-related genes (*atpD*, *glnⅡ* and *recA*), and two critical symbiosis-related genes (*nodC* and *nifH*) were described previously. The PCR products were purified and commercially sequenced by ABI 3730XL sequenator in Beijing, China. All obtained sequences were manually checked using Chromas Pro (Ver. 1. 7. 6;

Technelysium).

Phylogenetic analyses

The nucleotide sequences for all these genes were aligned and neighbour-joining (NJ) phylogenetic trees were constructed using the programme ClustalW integrated MEGA5 with a Kimura 2-parameter model. Genospecies were identified by calculating the ANI of taxonomic markers (*SMc00019*, *truA* and *thrA*) with a 96% threshold value for different species. Then, the degree of tree-like structures for the alleles of each locus and concatenated sequences were assessed to reveal potentially incompatible signals in the evolutionary history with split phylogenetic networks (1 000 bootstraps) using the SplitsTree4 programme. PhyML was used in combination with ModelTest 3.7 for maximum likelihood (ML) tree construction, and the Shimodaira-Hasegawa test was then performed to assess the topological consistency of trees inferred from different sets of sequence partitions as implemented in the PAUP* 4.0b1 programme. Recombination effects observed among these populations during the phylogenetic history were judged using ClonalFrame for five independent runs (500 000 burn-in iterations plus 1 000 000 sampling iterations for each run) based on three separate concatenated sequences.

Nucleotide polymorphism and population genetics analyses

Nucleotide polymorphisms, including the number of haplotypes (h), haplotype diversity (Hd), nucleotide diversity (π), number of synonymous substitutions per synonymous site (π_S), number of non-synonymous substitutions per non-synonymous site (π_N) and π_N/π_S ratio, were calculated by DnaSP v5. The average nucleotide divergence (Dxy) and number of migrants (Nm) were calculated using DnaSP v5 to infer the genetic divergence and gene flow among these populations. ClonalFrame was employed to calculate the two recombination rate statistics: r/m (the relative impact of recombination compared with that of the point mutation in the genetic diversification of the lineage) and ρ/θ (the relative frequency of the occurrence of recombination compared with that of the point mutation in the history of the lineage). DnaSP software was used to calculate the minimal recombination events (Rm) for persuasively estimating the recombination within the populations and between WFs and UAs ecological environments. Admixture levels of the bradyrhizobial genospecies inherited from the K 'ancestral' subpopulations were estimated using STRUCTURE with the admixture LOCPRIOR model.

Nucleotide sequence accession numbers

The 752 nucleotide sequences obtained in this study were deposited in the GenBank database under accession numbers (Tab. 4-1). The other remaining 78 nucleotide sequences of 13 type strains used in the phylogenetic trees were directly obtained from the GenBank database (Tab. 4-2).

Tab. 4-1 GenBank accession numbers of different genes of *Kummerowia* bradyrhizobia in this study

Strains	Accession numbers in GenBank							
	atpD	glnII	recA	SMc00019	thrA	truA	nifH	nodC
15640	KC356084	KC356176	KC356471	KC356571	KC356671	KC356771	KC356276	KC356374
15644	KC356085	KC356177	KC356472	KC356572	KC356672	KC356772	KC356277	KC356375
15647	KC356086	KC356178	KC356473	KC356573	KC356673	KC356773	KC356278	KC356376
15653	KC356087	KC356179	KC356474	KC356574	KC356674	KC356774	KC356279	KC356377
15658	KC356088	KC356180	KC356475	KC356575	KC35675	KC356775	KC356280	KC356378
15662	KC356089	KC356181	KC356476	KC356576	KC356676	KC356776	KC356281	KC356379
15732	KC356090	KC356182	KC356477	KC356577	KC356677	KC356777	KC356282	KC356380
101057	KC356159	KC356254	KC356549	KC356649	KC356749	KC356849	KC356352	KC356450
101058	KC356160	KC356255	KC356550	KC356650	KC356750	KC356850	KC356353	KC356451
101059	KC356161	KC356256	KC356551	KC356651	KC356751	KC356851	KC356354	KC356452
101060	KC356162	KC356257	KC356552	KC356652	KC356752	KC356852	KC356355	KC356453
101061	KC356163	KC356258	KC356553	KC356653	KC356753	KC356853	KC356356	KC356454
101062	KC356164	KC356259	KC356554	KC356654	KC356754	KC356854	KC356357	KC356455

Continued

Strains	Accession numbers in GenBank							
	atpD	glnII	recA	SMc00019	thrA	truA	nifH	nodC
101063	KC356165	KC356260	KC356555	KC356655	KC356755	KC356855	KC356358	KC356456
101064	KC356166	KC356261	KC356556	KC356656	KC356756	KC356856	KC356359	KC356457
101066	KC356168	KC356263	KC356558	KC356658	KC356758	KC356858	KC356361	KC356458
101067	KC356169	KC356264	KC356559	KC356659	KC356759	KC356859	KC356362	KC356459
101068	KC356170	KC356265	KC356560	KC356660	KC356760	KC356860	KC356363	KC356460
101069	KC356171	KC356266	KC356561	KC356661	KC356761	KC356861	KC356364	KC356461
101070	KF147151	KC356267	KC356562	KC356662	KC356762	KC356862	KC356365	KC356462
101071	KC356172	KC356268	KC356563	KC356663	KC356763	KC356863	KC356366	KC356463
101072	KC356173	KC356269	KC356564	KC356664	KC356764	KC356864	KC356367	KC356464
101073	KF147152	KC356270	KC356565	KC356665	KC356765	KC356865	KC356368	KC356465
101074	KC356174	KC356271	KC356566	KC356666	KC356766	KC356866	KC356369	KC356466
101075	KF147153	KC356272	KC356567	KC356667	KC356767	KC356867	KC356370	KC356467

Continued

Strains	Accession numbers in GenBank							
	atpD	glnII	recA	SMc00019	thrA	truA	nifH	nodC
101077	KC356175	KC356274	KC356569	KC356669	KC356769	KC356869	KC356372	KC356469
101078	KF147154	KC356275	KC356570	KC356670	KC356770	KC356870	KC356373	KC356470
25625	KC356097	KC356189	KC356484	KC356584	KC356684	KC356784	KC356289	KC356387
25626	KC356098	KC356190	KC356485	KC356585	KC356685	KC356785	KC356290	KC356388
25627	KC356099	KC356191	KC356486	KC356586	KC356686	KC356786	KC356291	KC356389
25628	KC356100	KC356192	KC356487	KC356587	KC356687	KC356787	KC356292	KC356390
25629	KC356101	KC356193	KC356488	KC356588	KC356688	KC356788	KC356293	KC356391
25630	KC356102	KC356194	KC356489	KC356589	KC356689	KC356789	KC356294	KC356392
25631	KC356103	KC356195	KC356490	KC356590	KC356690	KC356790	KC356295	KC356393
25632	KC356104	KC356196	KC356491	KC356591	KC356691	KC356791	KC356296	KC356394
25633	KC356105	KC356197	KC356492	KC356592	KC356692	KC356792	KC356297	KC356395
25634	KC356106	KC356198	KC356493	KC356593	KC356693	KC356793	KC356298	KC356396

Continued

Strains	Accession numbers in GenBank							
	atpD	glnII	recA	SMc00019	thrA	truA	nifH	nodC
25635	KC356107	KC356199	KC356494	KC356594	KC356694	KC356794	KC356299	KC356397
25636	KC356108	KC356200	KC356495	KC356595	KC356695	KC356795	KC356300	KC356398
25637	KC356109	KC356201	KC356496	KC356596	KC356696	KC356796	KC356301	KC356399
25638	KC356110	KC356202	KC356497	KC356597	KC356697	KC356797	KC356302	KC356400
25639	KC356111	KC356203	KC356498	KC356598	KC356698	KC356798	KC356303	KC356401
25640	KC356112	KC356204	KC356499	KC356599	KC356699	KC356799	KC356304	KC356402
25641	KC356113	KC356205	KC356500	KC356600	KC356700	KC356800	KC356305	KC356403
25642	KC356114	KC356206	KC356501	KC356601	KC356701	KC356801	KC356306	KC356404
25644	KC356116	KC356208	KC356503	KC356603	KC356703	KC356803	KC356307	KC356406
25645	KC356117	KC356209	KC356504	KC356604	KC356704	KC356804	KC356308	KC356407
25646	KC356118	KC356210	KC356505	KC356605	KC356705	KC356805	KC356309	KC356408
25647	KC356119	KC356211	KC356506	KC356606	KC356706	KC356806	KC356310	KC356409

Continued

Strains	Accession numbers in GenBank								
	atpD	glnII	recA	SMc00019	thrA	truA	nifH	nodC	
25648	KC356120	KC356212	KC356507	KC356607	KC356707	KC356807	KC356311	KC356410	
25649	KC356121	KC356213	KC356508	KC356608	KC356708	KC356808	KC356312	KC356411	
25650	KC356122	KC356214	KC356509	KC356609	KC356709	KC356809	KC356313	KC356412	
25651	KC356123	KC356215	KC356510	KC356610	KC356710	KC356810	KC356314	KC356413	
25652	KC356124	KC356216	KC356511	KC356611	KC356711	KC356811	KC356315	KC356414	
25653	KC356125	KC356217	KC356512	KC356612	KC356712	KC356812	KC356316	KC356415	
25654	KC356126	KC356218	KC356513	KC356613	KC356713	KC356813	KC356317	KC356416	
23371	KC356091	KC356183	KC356478	KC356578	KC356678	KC356778	KC356283	KC356381	
23372	KC356092	KC356184	KC356479	KC356579	KC356679	KC356779	KC356284	KC356382	
23373	KC356093	KC356185	KC356480	KC356580	KC356680	KC356780	KC356285	KC356383	
23374	KC356094	KC356186	KC356481	KC356581	KC356681	KC356781	KC356286	KC356384	
23375	KC356095	KC356187	KC356482	KC356582	KC356682	KC356782	KC356287	KC356385	

Continued

Strains	Accession numbers in GenBank							
	atpD	glnII	recA	SMc00019	thrA	truA	nifH	nodC
23376	KC356096	KC356188	KC356483	KC356583	KC356683	KC356783	KC356288	KC356386
23377	KF147155	KF147158	KF147160	KF147160	KF147161	KF147162	KF147163	KF147159
33006	KC356127	KC356219	KC356514	KC356614	KC356714	KC356814	KC356318	KC356417
33007	KC356128	KC356220	KC356515	KC356615	KC356715	KC356815	KC356319	KC356418
33037	KC356129	KC356221	KC356516	KC356616	KC356716	KC356816	KC356320	KC356419
33039	KC356130	KC356222	KC356517	KC356617	KC356717	KC356817	KC356321	KC356420
33100	KC356131	KC356223	KC356518	KC356618	KC356718	KC356818	KC356322	KC356421
33102	KC356132	KC356224	KC356519	KC356619	KC356719	KC356819	KC356323	KC356422
33118	KC356133	KC356225	KC356520	KC356620	KC356720	KC356820	KC356324	KC356423
33128	KF147156	KC356226	KC356521	KC356621	KC356721	KC356821	KC356325	KC356424
33158	KC356135	KC356228	KC356523	KC356623	KC356723	KC356823	KC356327	KC356426
43005	KC356143	KC356236	KC356531	KC356631	KC356731	KC356831	KC356335	KC356434

Continued

Strains	Accession numbers in GenBank							
	atpD	glnII	recA	SMc00019	thrA	truA	nifH	nodC
43048	KC356145	KC356238	KC356533	KC356633	KC356733	KC356833	KC356336	KC356435
43058	KC356146	KC356239	KC356534	KC356634	KC356734	KC356834	KC356337	KC356436
43078	KC356147	KC356240	KC356535	KC356635	KC356735	KC356835	KC356338	KC356437
43224	KF147157	KC356242	KC356537	KC356637	KC356737	KC356837	KC356340	KC356439
41003	KC356136	KC356229	KC356524	KC356624	KC356724	KC356824	KC356328	KC356427
41019	KC356137	KC356230	KC356525	KC356625	KC356725	KC356825	KC356329	KC356428
41032	KC356138	KC356231	KC356526	KC356626	KC356726	KC356826	KC356330	KC356429
41071	KC356139	KC356232	KC356527	KC356627	KC356727	KC356827	KC356331	KC356430
41136	KC356140	KC356233	KC356528	KC356628	KC356728	KC356828	KC356332	KC356431
41137	KC356141	KC356234	KC356529	KC356629	KC356729	KC356829	KC356333	KC356432
41152	KC356142	KC356235	KC356530	KC356630	KC356730	KC356830	KC356334	KC356433
61015	KC356149	KC356244	KC356539	KC356639	KC356739	KC356839	KC356342	KC356441

Continued

Strains	Accession numbers in GenBank							
	atpD	glnII	recA	SMc00019	thrA	truA	nifH	nodC
61017	KC356150	KC356245	KC356540	KC356640	KC356740	KC356840	KC356343	KC356442
61018	KC356151	KC356246	KC356541	KC356641	KC356741	KC356841	KC356344	KC356443
61022	KC356152	KC356247	KC356542	KC356642	KC356742	KC356842	KC356345	KC356444
61025	KC356153	KC356248	KC356543	KC356643	KC356743	KC356843	KC356346	KC356445
61035	KC356155	KC356250	KC356545	KC356645	KC356745	KC356845	KC356348	KC356446
61046	KC356156	KC356251	KC356546	KC356646	KC356746	KC356846	KC356349	KC356447
61051	KC356157	KC356252	KC356547	KC356647	KC356747	KC356847	KC356350	KC356448
61058	KC356158	KC356253	KC356548	KC356648	KC356748	KC356848	KC356351	KC356449

Tab. 4-2 GenBank accession numbers of different genes of the reference strains

Strains	GenBank Accession					
	atpD	glnII	recA	SMc00019	thrA	truA
B. betae LMG 21987[T]	FM253129	AM353733	AM353734	JX064202	JX064346	JX064274
B. canariense BTA-1[T]	FM253135	AY386765	AY591553	JX064204	JX064348	JX064276
B. daqingense CCBAU 15774[T]	HQ231289	HQ231301	HQ231270	JX064201	JX064345	JX064273
B. elkanii USDA 76[T]	AY386758	AY599117	AM591568	JX064205	JX064349	JX064277
B. huanghuaihaiense CCBAU 23303[T]	HQ231682	HQ231639	HQ231595	JX064203	JX064347	JX064275
B. iriomotense LMG 24129[T]	AB300994	AB300995	AB300996	JX064209	JX064353	JX064281
B. japonicum USDA 6[T]	AJ294388	AF169582	AM168341	JX064200	JX064344	JX064272
B. diazoefficiens USDA 110[T]	BA000040	BA000040	BA000040	BA000040	BA000040	BA000040
B. jicamae PAC 68[T]	FJ428211	FJ428204	HM047133	JX064207	JX064351	JX064279
B. lablabi CCBAU 23086[T]	GU433473	GU433498	GU433522	JX064208	JX064352	JX064280
B. liaoningense USDA 3622[T]	AY493450	AY494803	AY494833	JX064206	JX064350	JX064278
B. pachyrhizi PAC 48[T]	FJ428208	FJ428201	HM047130	JX064210	JX064354	JX064282
B. yuanmingense CCBAU 10071[T]	AY386760	AY386780	AY591566	JX064199	JX064343	JX064271

Results

Phylogenies of *Bradyrhizobium* nodulate *Kummerowia* in China

A total of 283 rhizobial isolates from earlier studies of *Kummerowia* rhizobia were subjected to sequence analyses of the gene *thrA*, one of the useful phylogenetic and taxonomic markers for rhizobia (not shown), and 94 representative strains were selected from further analysis (strains information showed in Tab. 4-3). The topological structures of the NJ and ML phylogenetic trees showed on concatenated sequences of three taxonomic markers (*SMc*00019, *truA* and *thrA*) using MEGA5 and PpyML, respectively, showed no significant difference. In view of this NJ phylogenetic tree, a total of seven genospecies were identified as follows: *B. diazoefficiens* (20 strains), *B. japonicum* (2 strains), *B. huanghuaihaiense* (1 strains), *B. liaoningense* (11 strains), *B. yuanmingense* (6 strains), *B. elkanii* (4 strains) and *B. pachyrhizi* (3 strains), according to the similarity percentage of interspecies (<94%) and intraspecies (>96%) ANI value boundaries. The ANI values of another seven genospecies were identified as new genospecies (designated as *Bradyrhizobium* sp. I to VII). The topological structures of the three taxonomic markers and the ecoregion-related genes showed consistency for most strains, although very few variations could not be well grouped in their own genospecies (Fig. 4-1 and 4-2). However, the phylogenetic tree was reconstructed for the concatenated sequences of the symbiotic genes of these strains, which were intermingled regardless of whether they were isolated from southern or northern China (Fig. 4-3). Interestingly, the low levels of biodiversity for strains isolated from UAs were grouped together and distributed into three new genospecies, with the exception of strain *B. yuanmingense* CCBAU 101063. In contrast, high biodiversity was observed in strains isolated from WFs and were separated in seven previously identified and four new genospecies.

Tab. 4-3 Total 94 representative strains isolated from *Kummerowia* used in this study

Strains (CCBAU No.)	Species	Collector	Longitude	Latitude	Geographic origin	Reference
15640	*B. diazoefficiens*	Hao Wang	127°20'02.45"	45°48'10.73"	Zhangguangcai Mountain, Bin County, Heilongjiang	Wang et al., 2009
15644	*B. diazoefficiens*	Hao Wang	127°20'02.45"	45°48'10.73"	Zhangguangcai Mountain, Bin County, Heilongjiang	Wang et al., 2009
15647	*Bradyrhizobium* sp. II	Hao Wang	127°20'02.45"	45°48'10.73"	Zhangguangcai Mountain, Bin County, Heilongjiang	Wang et al., 2009
15653	*Bradyrhizobium* sp. I	Hao Wang	127°20'02.45"	45°48'10.73"	Zhangguangcai Mountain, Bin County, Heilongjiang	Wang et al., 2009
15658	*Bradyrhizobium* sp. II	Hao Wang	127°20'02.45"	45°48'10.73"	Zhangguangcai Mountain, Bin County, Heilongjiang	Wang et al., 2009
15662	*Bradyrhizobium* sp. I	Hao Wang	127°20'02.45"	45°48'10.73"	Zhangguangcai Mountain, Bin County, Heilongjiang	Wang et al., 2009

Continued

Strains (CCBAU No.)	Species	Collector	Longitude	Latitude	Geographic origin	Reference
15732	Bradyrhizobium sp. I	Hao Wang	127°20′02.45″	45°48′10.73″	Zhangguangcai Mountain, Bin County, Heilongjiang	Wang et al., 2009
23371	B. elkanii	Tianyan Liu	118°34′33.90″	31°41′53.24″	Maanshan City, Anhui	This study
23372	Bradyrhizobium sp. III	Tianyan Liu	118°34′33.90″	31°41′53.24″	Maanshan City, Anhui	This study
23373	B. diazoefficiens	Tianyan Liu	118°34′33.90″	31°41′53.24″	Maanshan City, Anhui	This study
23374	B. diazoefficiens	Tianyan Liu	118°34′33.90″	31°41′53.24″	Maanshan City, Anhui	This study
23375	B. diazoefficiens	Tianyan Liu	118°34′33.90″	31°41′53.24″	Maanshan City, Anhui	This study
23376	B. diazoefficiens	Tianyan Liu	118°34′33.90″	31°41′53.24″	Maanshan City, Anhui	This study
23377	B. diazoefficiens	Tianyan Liu	118°34′33.90″	31°41′53.24″	Maanshan City, Anhui	This study
25625	B. diazoefficiens	Tianyan Liu	118°47′33″	35°48′12″	Anzhuang Town, Juxian County, Rizhao City, Shandong	This study

Continued

Strains (CCBAU No.)	Species	Collector	Longitude	Latitude	Geographic origin	Reference
25626	B. liaoningense	Tianyan Liu	118°47′33″	35°48′12″	Anzhuang Town, Juxian County, Rizhao City, Shandong	This study
25627	B. liaoningense	Tianyan Liu	118°47′33″	35°48′12″	Anzhuang Town, Juxian County, Rizhao City, Shandong	This study
25628	Bradyrhizobium sp. I	Tianyan Liu	118°47′33″	35°48′12″	Anzhuang Town, Juxian County, Rizhao City, Shandong	This study
25629	B. liaoningense	Tianyan Liu	118°47′33″	35°48′12″	Anzhuang Town, Juxian County, Rizhao City, Shandong	This study

Continued

Strains (CCBAU No.)	Species	Collector	Longitude	Latitude	Geographic origin	Reference
25630	B. diazoefficiens	Tianyan Liu	118°47′33″	35°48′12″	Anzhuang Town, Juxian County, Rizhao City, Shandong	This study
25631	B. liaoningense	Tianyan Liu	118°47′33″	35°48′12″	Anzhuang Town, Juxian County, Rizhao City, Shandong	This study
25632	B. diazoefficiens	Tianyan Liu	118°47′33″	35°48′12″	Anzhuang Town, Juxian County, Rizhao City, Shandong	This study
25633	Bradyrhizobium sp. I	Tianyan Liu	118°43′16″	35°46′32″	Longtanguanzhuang Village, Juxian County, Rizhao City, Shandong	This study

Continued

Strains (CCBAU No.)	Species	Collector	Longitude	Latitude	Geographic origin	Reference
25634	B. diazoefficiens	Tianyan Liu	118°43′16″	35°46′32″	Longtanguanzhuang Village, Juxian County, Rizhao City, Shandong	This study
25635	B. liaoningense	Tianyan Liu	118°43′16″	35°46′32″	Longtanguanzhuang Village, Juxian County, Rizhao City, Shandong	This study
25636	B. liaoningense	Tianyan Liu	118°43′16″	35°46′32″	Longtanguanzhuang Village, Juxian County, Rizhao City, Shandong	This study
25637	Bradyrhizobium sp. I	Tianyan Liu	118°43′16″	35°46′32″	Longtanguanzhuang Village, Juxian County, Rizhao City, Shandong	This study
25638	Bradyrhizobium sp. I	Tianyan Liu	118°52′18″	35°05′04″	Qishan Town, Juxian County, Rizhao City, Shandong	This study

Continued

Strains (CCBAU No.)	Species	Collector	Longitude	Latitude	Geographic origin	Reference
25639	B. liaoningense	Tianyan Liu	118°52′18″	35°05′04″	Qishan Town, Juxian County, Rizhao City, Shandong	This study
25640	B. diazoefficiens	Tianyan Liu	118°52′18″	35°05′04″	Qishan Town, Juxian County, Rizhao City, Shandong	This study
25641	B. liaoningense	Tianyan Liu	118°52′18″	35°05′04″	Qishan Town, Juxian County, Rizhao City, Shandong	This study
25642	Bradyrhizobium sp. II	Tianyan Liu	115°38′47″	33°50′32″	Dazhuangpo, Qishan Town, Juxian County, Rizhao City, Shandong	This study
25644	B. elkanii	Tianyan Liu	115°38′47″	33°50′32″	Dazhuangpo Village, Qishan Town, Juxian County, Rizhao City, Shandong	This study

Continued

Strains (CCBAU No.)	Species	Collector	Longitude	Latitude	Geographic origin	Reference
25645	B. pachyrhizi	Tianyan Liu	115°38'47"	33°50'32"	Dazhuangpo Village, Qishan Town, Juxian County, Rizhao City, Shandong	This study
25646	B. pachyrhizi	Tianyan Liu	115°38'47"	33°50'32"	Dazhuangpo Village, Qishan Town, Juxian County, Rizhao City, Shandong	This study
25647	B. pachyrhizi	Tianyan Liu	115°38'47"	33°50'32"	Dazhuangpo Village, Qishan Town, Juxian County, Rizhao City, Shandong	This study
25648	B. huanghuaihaiense	Tianyan Liu	115°38'47"	33°50'32"	Dazhuangpo Village, Qishan Town, Juxian County, Rizhao City, Shandong	This study

Continued

Strains (CCBAU No.)	Species	Collector	Longitude	Latitude	Geographic origin	Reference
25649	*B. elkanii*	Tianyan Liu	115°38′47″	33°50′32″	Dazhuangpo Village, Qishan Town, Juxian County, Rizhao City, Shandong	This study
25650	*Bradyrhizobium* sp. I	Tianyan Liu	115°38′47″	33°50′32″	Dazhuangpo Village, Qishan Town, Juxian County, Rizhao City, Shandong	This study
25651	*B. diazoefficiens*	Tianyan Liu	118°51′09″	35°47′30″	Side of Qingfengling Reservoir, Juxian County, Rizhao City, Shandong	This study
25652	*B. diazoefficiens*	Tianyan Liu	118°51′09″	35°47′30″	Side of Qingfengling Reservoir, Juxian County, Rizhao City, Shandong	This study

Continued

Strains (CCBAU No.)	Species	Collector	Longitude	Latitude	Geographic origin	Reference
25653	B. diazoefficiens	Tianyan Liu	118°51′09″	35°47′30″	Side of Qingfengling Reservoir, Juxian County, Rizhao City, Shandong	This study
25654	B. diazoefficiens	Tianyan Liu	118°51′09″	35°47′30″	Side of Qingfengling Reservoir, Juxian County, Rizhao City, Shandong	This study
33006	B. diazoefficiens	Dongxu Lin	114°55′11″	28°14′11″	Suburb of Shanggao County, Jiangxi	Lin et al., 2007
33007	B. diazoefficiens	Dongxu Lin	114°55′11″	28°14′11″	Suburb of Shanggao County, Jiangxi	Lin et al., 2007
33037	B. diazoefficiens	Dongxu Lin	116°37′17″	28°13′46″	Dongxiang District, Fuzhou City, Jiangxi	Lin et al., 2007

Continued

Strains (CCBAU No.)	Species	Collector	Longitude	Latitude	Geographic origin	Reference
33039	Bradyrhizobium sp. IV	Dongxu Lin	116°37′17″	28°13′46″	Dongxiang District, Fuzhou City, Jiangxi	Lin et al., 2007
33100	B. yuanmingense	Dongxu Lin	116°31′16″	27°13′18″	Suburb of Nanfeng County, Jiangxi	Lin et al., 2007
33102	B. yuanmingense	Dongxu Lin	116°31′16″	27°13′18″	Suburb of Nanfeng County, Jiangxi	Lin et al., 2007
33118	B. liaoningense	Dongxu Lin	114°58′04″	27°06′32″	Suburb of Ji'an City, Jiangxi	Lin et al., 2007
33128	B. japonicum	Dongxu Lin	116°02′32″	29°27′08″	Suburb of Nankang District, Jiangxi	Lin et al., 2007
33146	B. elkanii	Dongxu Lin	114°56′43″	26°50′07″	Suburb of Taihe County, Jiangxi	Lin et al., 2007
33158	B. liaoningense	Dongxu Lin	115°00′24″	25°51′49″	Suburb of Ganxian District, Jiangxi	Lin et al., 2007

Continued

Strains (CCBAU No.)	Species	Collector	Longitude	Latitude	Geographic origin	Reference
41003	*Bradyrhizobium* sp. IV	Dongxu Lin	112°56′00″	28°13′54″	Campus of Central South University of Forestry and Technology, ChangSha City, Hunan	Lin et al., 2007
41019	*Bradyrhizobium* sp. IV	Dongxu Lin	112°36′33″	27°23′24″	Hengshan Mountain, Hengyan City, Hunan	Lin et al., 2007
41032	*Bradyrhizobium* sp. IV	Dongxu Lin	112°36′33″	27°23′24″	Hengshan Mountain, Hengyan City, Hunan	Lin et al., 2007
41071	*Bradyrhizobium* sp. IV	Dongxu Lin	113°38′16″	28°10′02″	Liuyang City, Hunan	Lin et al., 2007
41136	*Bradyrhizobium* sp. IV	Dongxu Lin	110°37′38″	26°43′49″	Wugang City, Hunan	Lin et al., 2007
41137	*Bradyrhizobium* sp. IV	Dongxu Lin	110°34′17″	27°03′50″	Dongkou County, Hunan	Lin et al., 2007
41152	*Bradyrhizobium* sp. IV	Dongxu Lin	109°56′46″	28°37′14″	Guzhang County, Hunan	Lin et al., 2007
43005	*B. yuanmingense*	Dongxu Lin	112°15′30″	29°14′53″	Daoguan River, Xinzhou District, Hubei	Lin et al., 2007

Continued

Strains (CCBAU No.)	Species	Collector	Longitude	Latitude	Geographic origin	Reference
43048	B. liaoningense	Dongxu Lin	113°48'41"	29°43'49"	Dazeping Town, Tongcheng County, Hubei	Lin et al., 2007
43058	Bradyrhizobium sp. I	Dongxu Lin	112°04'09"	32°0'24"	Scenic spot of Longzhong, Xiangyang City, Hubei	Lin et al., 2007
43078	Bradyrhizobium sp. I	Dongxu Lin	113°17'36"	31°51'21"	Suburb of Suixian County, Hubei	Lin et al., 2007
43224	B. japonicum	Dongxu Lin	111°16'51"	30°41'40"	Suburb of Yichang City, Hubei	Lin et al., 2007
61015	Bradyrhizobium sp. III	Qiang Chen	104°40'35"	31°28'11"	Mianyang City, Sichuang	Chen, 2004
61017	Bradyrhizobium sp. I	Qiang Chen	107°05'58"	31°33'46"	Pingchang County, Sichuan	Chen, 2004
61018	Bradyrhizobium sp. I	Qiang Chen	107°05'58"	31°33'46"	Pingchang County, Sichuan	Chen, 2004
61022	Bradyrhizobium sp. I	Qiang Chen	107°05'58"	31°33'46"	Pingchang County, Sichuan	Chen, 2004
61025	Bradyrhizobium sp. I	Qiang Chen	107°05'58"	31°33'46"	Pingchang County, Sichuan	Chen, 2004

Continued

Strains (CCBAU No.)	Species	Collector	Longitude	Latitude	Geographic origin	Reference
61035	B. yuanmingense	Qiang Chen	104°40'35"	31°28'11"	Mianyang City, Sichuang	Chen, 2004
61046	Bradyrhizobium sp. III	Qiang Chen	104°40'35"	31°28'11"	Mianyang City, Sichuang	Chen, 2004
61051	B. diazoefficiens	Qiang Chen	107°05'58"	31°33'46"	Pingchang County, Sichuan	Chen, 2004
61058	B. yuanmingense	Qiang Chen	107°05'58"	31°33'46"	Pingchang County, Sichuan	Chen, 2004
101057	Bradyrhizobium sp. VI	Tianyan Liu	116°16'27.55"	40°01'28.23"	West campus of CAU, Haidian District, Beijing	This study
101058	Bradyrhizobium sp. VII	Tianyan Liu	116°16'27.55"	40°01'28.23"	West campus of CAU, Haidian District, Beijing	This study
101059	Bradyrhizobium sp. V	Tianyan Liu	116°16'27.55"	40°01'28.23"	West campus of CAU, Haidian District, Beijing	This study
101060	Bradyrhizobium sp. VI	Tianyan Liu	116°16'27.55"	40°01'28.23"	West campus of CAU, Haidian District, Beijing	This study

Continued

Strains (CCBAU No.)	Species	Collector	Longitude	Latitude	Geographic origin	Reference
101061	*Bradyrhizobium* sp. VI	Tianyan Liu	116°16′27.55″	40°01′28.23″	West campus of CAU, Haidian District, Beijing	This study
101062	*Bradyrhizobium* sp. VII	Tianyan Liu	116°16′27.55″	40°01′28.23″	West campus of CAU, Haidian District, Beijing	This study
101063	*B. yuanmingense*	Tianyan Liu	116°16′27.55″	40°01′28.23″	West campus of CAU, Haidian District, Beijing	This study
101064	*Bradyrhizobium* sp. VII	Tianyan Liu	116°16′27.55″	40°01′28.23″	West campus of CAU, Haidian District, Beijing	This study
101066	*Bradyrhizobium* sp. VII	Tianyan Liu	116°16′27.55″	40°01′28.23″	West campus of CAU, Haidian District, Beijing	This study
101067	*Bradyrhizobium* sp. VI	Tianyan Liu	116°16′27.55″	40°01′28.23″	West campus of CAU, Haidian District, Beijing	This study

Continued

Strains (CCBAU No.)	Species	Collector	Longitude	Latitude	Geographic origin	Reference
101068	*Bradyrhizobium* sp. VII	Tianyan Liu	116°16′27.55″	40°01′28.23″	West campus of CAU, Haidian District, Beijing	This study
101069	*Bradyrhizobium* sp. VII	Tianyan Liu	116°16′27.55″	40°01′28.23″	West campus of CAU, Haidian District, Beijing	This study
101070	*Bradyrhizobium* sp. V	Tianyan Liu	116°16′27.55″	40°01′28.23″	West campus of CAU, Haidian District, Beijing	This study
101071	*Bradyrhizobium* sp. VII	Tianyan Liu	116°16′27.55″	40°01′28.23″	West campus of CAU, Haidian District, Beijing	This study
101072	*Bradyrhizobium* sp. VII	Tianyan Liu	116°16′27.55″	40°01′28.23″	West campus of CAU, Haidian District, Beijing	This study
101073	*Bradyrhizobium* sp. V	Tianyan Liu	116°16′27.55″	40°01′28.23″	West campus of CAU, Haidian District, Beijing	This study

Continued

Strains (CCBAU No.)	Species	Collector	Longitude	Latitude	Geographic origin	Reference
101074	*Bradyrhizobium* sp. VII	Tianyan Liu	116°16′27.55″	40°01′28.23″	West campus of CAU, Haidian District, Beijing	This study
101075	*Bradyrhizobium* sp. V	Tianyan Liu	116°16′27.55″	40°01′28.23″	West campus of CAU, Haidian District, Beijing	This study
101077	*Bradyrhizobium* sp. VI	Tianyan Liu	116°16′27.55″	40°01′28.23″	West campus of CAU, Haidian District, Beijing	This study
101078	*Bradyrhizobium* sp. V	Tianyan Liu	116°16′27.55″	40°01′28.23″	West campus of CAU, Haidian District, Beijing	This study

Note: CCBAU, Culture Collection of Beijing Agricultural University

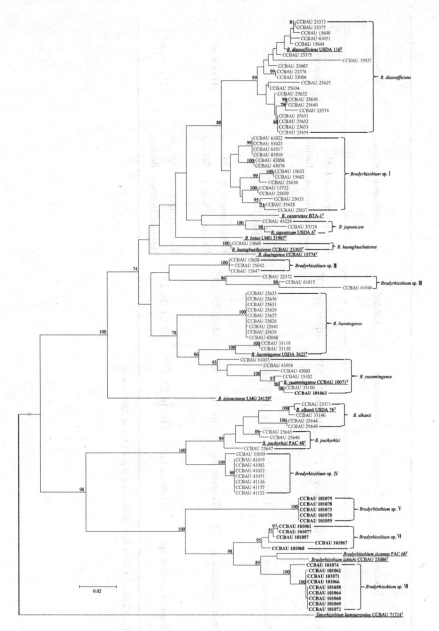

Fig. 4-1　NJ tree constructed based upon the concatenated sequences of three taxonomic markers. Three genes *SMc*00019, *truA*, and *thrA* were used as taxonomic markers. Bootstrap values greater than 70% are indicated at the branch points. Type (superscript T) strains underlined are used for references. The scale bar represents 1% nucleotide substitutions

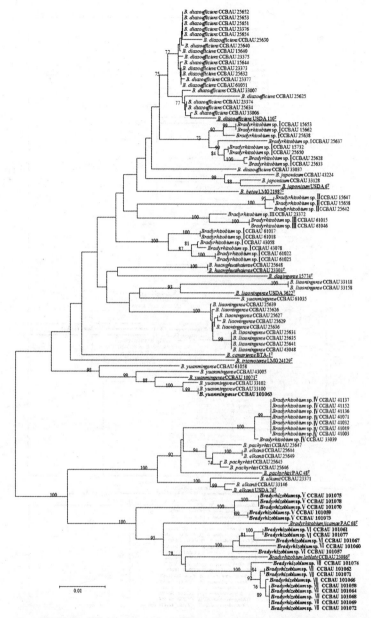

Fig. 4-2 NJ tree constructed based upon the concatenated sequences of ecoregion-related genes. Three ecoregion-related genes *atpD*, *glnII*, and *recA* were used. Bootstrap values greater than 70% are indicated at the branch points. Novel genospecies (*Bradyrhizobium* sp. I - VII) and type strains are boldfaced. The scale bar represents 1% nucleotide substitutions

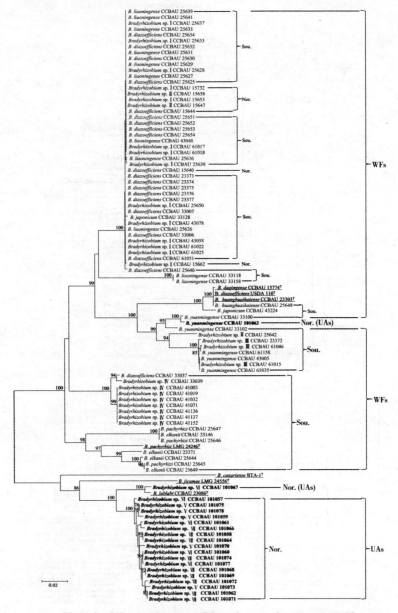

Fig. 4-3 NJ tree constructed based upon the concatenated sequences of symbiosis-related genes. Two symbiosis-related genes (*nodC* and *nifH*) were used. Bootstrap values greater than 70% are indicated at the branch points. Sou. or Nor. indicates that the strains were isolated from southern or northern ecoregions of China, respectively. WF, wild field. UA, urban area. Type (superscript T) strains underlined are used for references. The scale bar represents 1% nucleotide substitutions

Nucleotide diversity inferred from different genes and ecological environments

Nucleotide polymorphisms of the 94 representative strains for the three kinds of genes (taxonomic gene markers, ecoregion- and symbiosis-related genes) were calculated and the various parameters are listed in Tab. 4-4. The sequence type and haplotype diversity (Hd) of symbiosis-related genes were obviously lower than those of the other genes. The nucleotide diversity (π) values showed little difference among the three taxonomic markers or ecoregion-related genes, but the largest deviation of π value (0.118 86 for *nodC* and 0.066 46 for *nifH*) was found for symbiosis-related genes (Tab. 4-4); this was due to the large difference between their corresponding π_N (0.032 99 for *nodC* and 0.006 56 for *nifH*), ard not their π_S values (0.356 72 for *nodC* and 0.344 85 for *nifH*), as shown in Tab. 4-4. In addition, the strains isolated from UAs had lower π, π_N and π_S values compared with strains from WFs for all the test genes (Tab. 4-5). Significant differences ($P < 0.05$) in π value were not only found between WFs and UAs, but also between northern and southern China for all of the genes. The highest π values were presented in WFs (0.092 54) and southern China (0.094 96) for three taxonomic markers and the lowest π values were presented in UAs (0.028 27) and northern China (0.000 70) for symbiosis-related genes (Tab. 4-5).

The π_N/π_S values showed large differences (0.019 for *nifH* to 0.181 9 for *SMc00019*) for the test genes (Tab. 4-4). The π_N/π_S values were higher in WFs than in UAs for the three taxonomic markers and ecoregion-related genes; however, lower for the symbiosis-related genes. The π_N/π_S values were also higher for strains in southern China than for those in northern China for all the genes (Tab. 4-5).

Tab. 4-4 Molecular diversity for the three kinds of genes

Genes	Length/bp	S	Eta	h/Hd	π	π_S	π_N	π_N/π_S
Taxonomic markers								
smc00019	288	96	136	36/0.965	0.105 69	0.349 56	0.056 77	0.162 40
truA	450	194	273	46/0.971	0.135 87	0.442 12	0.080 44	0.181 94
thrA	474	159	217	44/0.972	0.104 19	0.484 01	0.023 80	0.049 17
Concatenate	1 212	449	626	57/0.981	0.116 31	0.431 88	0.052 08	0.120 59
Avg	404	149.7	208.7	42/0.969	0.115 25	0.425 23	0.053 67	0.131 17
Ecoregion-related genes								
recA	423	113	155	42/0.967	0.068 40	0.280 47	0.009 74	0.034 72
glnII	516	135	164	41/0.965	0.073 97	0.259 54	0.033 26	0.128 15
atpD	432	89	108	40/0.946	0.049 27	0.185 18	0.011 99	0.064 75
Concatenate	1 371	337	427	60/0.985	0.064 47	0.239 78	0.019 35	0.080 70
Avg	457	112.3	142.3	41/0.959	0.063 88	0.241 73	0.018 33	0.075 87
Symbiosis-related genes								
nifH	633	177	236	28/0.789	0.066 46	0.344 85	0.006 56	0.019 02

Continued

Genes	Length/bp	S	Eta	h/Hd	π	π_S	π_N	π_N/π_S
nodC	450	187	263	30/0.886	0.118 86	0.356 72	0.032 99	0.092 48
Concatenate	1 083	364	499	39/0.900	0.088 23	0.455 50	0.017 70	0.038 86
Avg	541.5	182	249.5	29/0.838	0.092 66	0.350 79	0.019 78	0.055 75

Note: S, segregating sites or number of polymorphic (segregating) sites; Eta, total number of mutations; h, haplotype number; Hd, haplotype diversity; π, average number of nucleotide differences per site between two sequences; π_S, nucleotide diversity for synonymous substitutions; π_N, nucleotide diversity for nonsynonymous substitutions

Tab. 4-5 Nucleotide polymorphism for the three kinds of genes in *Kummerowia* bradyrhizobia

Area (No. of strains)	Length/bp	S	Eta	h/Hd	π	π_S	π_N	π_N/π_S
Concatenated taxonomic markers								
WFs(74)	1 212	396	512	47/0.977	0.092 54	0.340 82	0.036 68	0.107 62
UAs(20)	1 212	284	331	10/0.884	0.071 03	0.289 39	0.021 00	0.072 57
Nor. (27)	1 212	129	138	5/0.905	0.051 98	0.206 28	0.010 25	0.049 69
Sou. (67)	1 212	395	507	43/0.973	0.094 96	0.348 00	0.038 57	0.110 83

Continued

Area (No. of strains)	Length/bp	S	Eta	h/Hd	π	π_S	π_N	π_N/π_S
Concatenated ecoregion-related genes								
WFs (74)	1 371	300	366	50/0.982	0.054 94	0.202 89	0.015 80	0.077 87
UAs (20)	1 371	183	199	11/0.916	0.038 16	0.155 89	0.005 68	0.036 44
Nor. (27)	1 371	83	88	5/0.905	0.030 22	0.108 19	0.007 52	0.069 51
Sou. (67)	1 371	300	366	46/0.979	0.056 37	0.208 34	0.016 32	0.078 33
Concatenated symbiosis-related genes								
WFs (74)	1 083	320	403	24/0.839	0.057 02	0.262 90	0.009 36	0.035 60
UAs (20)	1 083	217	236	15/0.974	0.028 27	0.141 34	0.006 98	0.049 38
Nor. (27)	1 083	2	2	3/0.524	0.000 70	0.002 78	0	0
Sou. (67)	1 083	320	402	23/0.860	0.061 55	0.284 88	0.010 13	0.035 56

Note: S, segregating sites or number of polymorphic (segregating) sites; Eta, total number of mutations; h, haplotype number; Hd, haplotype diversity; π, average number of nucleotide differences per site between two sequences; π_S, nucleotide diversity for synonymous substitutions; π_N, nucleotide diversity for nonsynonymous substitutions; WFs, wild fields; UAs, urban areas

Gene flow and genetic divergence among *Bradyrhizobium* of *Kummerowia* in two segregated ecoregions

Three different NeighborNet trees are shown here to analyse the gene flow and genetic exchange among 94 representative strains isolated from different ecoregions based on the three taxonomic markers, ecoregion- and symbiosis-related genes. Interweaves of the test genes were found clearly among some strains. Genetic exchanges and gene flows occurred more frequently among strains in the NeighborNet trees of the three taxonomic markers [Fig. 4-4(a)] and symbiosis-related genes [Fig. 4-4(b)] than ecoregion-related genes (Fig. 4-5). Furthermore, the gene flow among strains isolated both from southern and northern China was frequent but seldom occurred between the strains from WFs and UAs for the three kinds of test genes (Fig. 4-4 and 4-5).

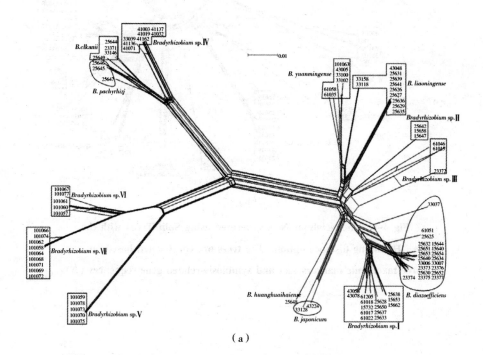

(a)

146 Geographic Distribution and Molecular Evolutionary Mechanism for *Mesorhizobium* Strains in Desert Soil

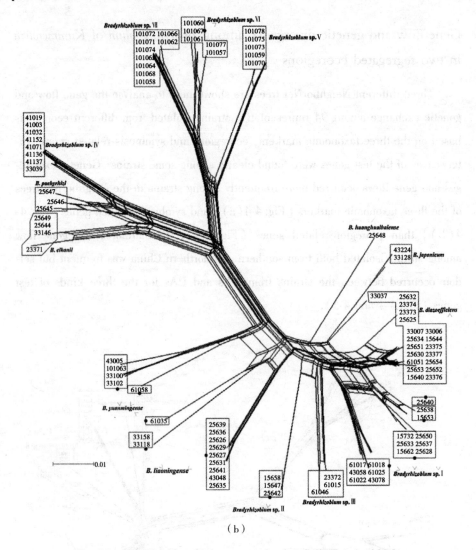

(b)

Fig. 4-4 The Neighbor-Nets generated using SplitsTree4 with the Hamming distance option. The trees are constructed based on three taxonomic markers (a) and symbiosis-related gene sequences (b)

Chapter IV Genetic divergence among *Bradyrhizobium* strains nodulating wild and cultivated *Kummerowia* spp. in China

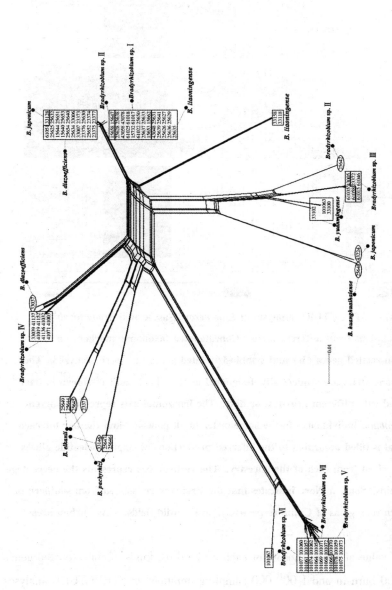

Fig. 4-5 The Neighbor-Nets generated using SplitsTree4 with the Hamming distance option. The trees are constructed based on the concatenated ecoregion-related genes (*atpD-glnII -recA*)

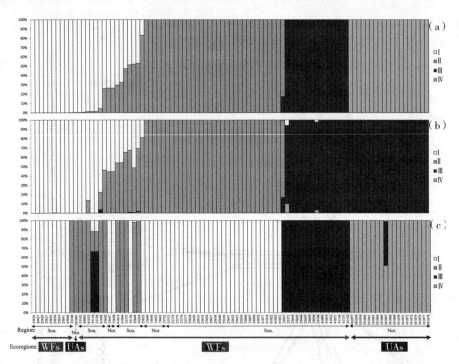

Fig. 4-6 STRUCTURE analyses of *Kummerowia*-associated *Bradyrhizobium* populations from different areas. Concatenated taxonomic markers (a), ecoregion-related genes (b) and symbiosis-related genes (c) were analyzed. The inferred ancestries are sequentially designated as I, II, III and IV shown in bars and filled with different grayness or dots. The horizontal axis represents current *Bradyrhizobium* individuals (in the same order in all panels), and the bar for each individual is filled according to the inferred proportions of single-nucleotide alleles that were derived from each of the ancestry. The vertical axis represents the percentage of the strains. Sou. or Nor. indicates that the strains were isolated from southern or northern ecoregions of China respectively. WFs, wild fields. UAs, urban areas

The K value of 4 was chosen for each of the three kinds of concatenated genes with 100 000 burn-in and 1 000 000 sampling iterations in STRUCTURE analyses and the strains were passively arranged in order on the basis of genetic exchange (Fig. 4-6). Four lineages (I to IV) showing clear and similar structure patterns were observed in lineage I and II for the three taxonomic markers and ecoregion-related genes, and in lineage III and IV for the three taxonomic markers and symbiosis-related genes, respectively. In addition, the gene flow between lineages I and II occurred frequently for the three taxonomic markers and ecoregion-related genes. In contrast, all the isolates obtained from UAs were located in one of the lineages that seldom intermingled with the isolates from WFs for all kinds of genes.

As shown in Tab. 4-6 and 4-7, the highest genetic distances (Dxy) were detected and the p-values for the Dxy comparison pairs were statistically significant ($P < 0.05$) between WFs and UAs for the three taxonomic markers, ecoregion- and symbiosis-related genes (0.165 49, 0.085 24 and 0.152 88, respectively) and between southern and northern China (0.049 31 and 0.038 23 for the three taxonomic markers and ecoregion-related genes, respectively). However, no significant difference was found for the symbiosis-related genes (0.038 23) between the pairs of isolates obtained from southern and northern China ($P < 0.05$). Moreover, the Dxy values between WFs and UAs (0.165 49, 0.085 24 and 0.152 88 for the three taxonomic markers, ecoregion- and symbiosis-related genes, respectively) were much higher than the mean π value (0.081 79, 0.046 55 and 0.042 65 for the taxonomic markers, ecoregion- and symbiosis-related genes, respectively) (Tab. 4-6). Nevertheless, the Dxy values between southern and northern China (0.082 96, 0.049 31 and 0.038 23 for the taxonomic markers, ecoregion- and symbiosis-related genes, respectively) were close to the mean π value (0.073 47, 0.043 30 and 0.031 13 for taxonomic markers, ecoregion- and symbiosis-related genes, respectively) (Tab. 4-7).

Tab. 4-6 Genetic divergence (presented as Dxy) and gene flow (presented as Nm) in the *Kummerowia* bradyrhizobia from WFs and UAs

	Dxy	Nm
Taxonomic markers		
WF (74)	—	0.49
UA (20)	0.165 49[e]	—
Ecoregion-related genes		
WF (74)	—	0.6
UA (20)	0.085 24[e]	—
Symbiosis-related genes		
WF (74)	—	0.19
UA (20)	0.152 88[d]	—

Note: Number of gene flow (Nm) and average nucleotide divergence between groups (Dxy) are shown in the upper and lower triangular of the table.

Statistical difference letter is marked as superscript after each number in the lower triangular of the table: b, $0.01 < P < 0.05$; c, $0.001 < P < 0.01$; d, $0.0001 < P < 0.001$; e, non-significant

Tab. 4-7 Genetic divergence (presented as Dxy) and gene flow (presented as Nm) in the *Kummerowia* bradyrhizobia from southern and northern China

	Dxy	Nm
Taxonomic markers		
Nor. (27)	—	3.87
Sou. (67)	0.082 96[b]	—
Ecoregion-related genes		
Nor. (27)	—	3.6
Sou. (67)	0.049 31[b]	—

Continued

	Dxy	Nm
Symbiosis-related genes		
Nor. (27)	—	2.19
Sou. (67)	0.038 23^e	—

Note: Sou. or Nor. means the strains isolated from southern or northern ecoregions of China respectively. Number of gene flow (Nm) and average nucleotide divergence between groups (Dxy) are shown in the upper and lower triangular of the table.

Statistical difference letter is marked as superscript after each number in the lower triangular of the table: b, $0.01 < P < 0.05$; c, $0.001 < P < 0.01$; d, $0.000\ 1 < P < 0.001$; e, non-significant

Recombination and evolutionary lineages of *Bradyrhizobium* of *Kummerowia* in two segregated ecoregions

A total of 118, 93 and 82 putative recombination events (Rm) were inferred for the concatenated taxonomic markers, ecoregion- and symbiosis-related genes, respectively, with DnaSP software (Tab. 4-8). The symbiosis-related genes had the lower Rm value than other genes and lower levels of recombination rates in evolutionary history.

Recombination rates in evolutionary history were further assessed by calculating the r/m and ρ/θ ratios using the ClonalFrame programme. The relative frequency of the occurrence of recombination (ρ/θ) was less than 1.0 for all of the test genes compared with that of point mutation in the history of the lineage (Tab. 4-8). Howerer, the relative impact of recombination, compared with that of the point mutation (r/m), was greater than 1.0 for the taxonomic markers, ecoregion-related genes other than symbiosis-related genes. In addition, the Rm, ρ/θ and r/m values of *Kummerowia* bradyrhizobia isolated from UAs were much lower than those isolated from WFs for all of the test genes.

The topological structures were significantly incongruent with the phylogeny of the inferred species for the test genes other than *thrA* compared to three taxonomic markers *SMc*00019-*truA*-*thrA* ($P < 0.05$) in the Shimodaira-Hasegawa (SH) test by using the ModelTest 3.7 programme integrated in PAUP software (Tab. 4-9).

Tab. 4-8 Recombination analysis by DnaSP and ClonalFrame

Genes	Length/bp	Rm	r/m	ρ/θ
Concatenated taxonomic markers	1 212	118	1.39	0.29
Wild fields (WFs)	1 212	102	0.72	0.12
Urban areas (UAs)	1 212	40	0.62	0.07
Concatenated ecoregion-related genes	1 371	93	2.52	0.67
Wild fields (WFs)	1 371	85	2.30	0.54
Urban areas (UAs)	1 371	18	0.39	0.062
Concatenated symbiosis-related genes	1 083	82	0.17	0.01
Wild fields (WFs)	1 083	61	0.047	0.002 9
Urban areas (UAs)	1 083	10	0.012	0.000 17

Note: Rm, observed minimum number of recombination events; r/m, the relative impact of recombination compared with that of point mutation in the genetic diversification of the lineage; ρ/θ, the relative frequency of the occurrence of recombination compared with that of point mutation in the history of the lineage

Tab. 4-9 Shimodaira-Hasegawa test of each gene locus in comparison with the the concatenated taxonomic markers

Genes	-lnL	Diff -lnL	P
*SMc*00019	6 974.08	7 92.37	0.047*
truA	9 083.87	835.14	0.006*
thrA	3 274.31	475.61	0.146
atpD	8 945.36	901.65	0.016*
glnII	7 744.24	766.30	0.004*
recA	8 706.46	824.26	0.021*
nodC	9 012.51	998.27	0.001*

Continued

Genes	-lnL	Diff -lnL	P
nifH	8 090.73	1 104.19	0.000*

Note: -lnL, negative log-likelihood value for the constrained topology; Diff-lnL, score difference between the nonconstrained and constrained trees; P, significance of the difference in-lnL scores calculated based on the constrained and nonconstrained trees as assessed by the SH test; *, $P < 0.05$

Discussion

Kummerowia, a legume, usually grows wildly in nature; however, human has recently started cultivating it. The bacterial genus *Bradyrhizobium* is *Kummerowia*'s main symbiotic species and has high nitrogen fixation abilities. *Bradyrhizobium* can form nodules on the roots of *Kummerowia* in the soil. In this study, the evolutionary relationship of *Bradyrhizobium* was systematically analysed based on three taxonomic markers, three ecoregion-related genes and two symbiosis-related genes. The results showed that low levels of biodiversity, infrequent gene exchange, infrequent recombination and independent evolutionary lineage of *Bradyrhizobium* strains nodulating *Kummerowia* gradually evolved and emerged with urban segregation.

Low level of biodiversity for *Bradyrhizobium* strains nodulating *Kummerowia* with urban segregation

The *Bradyrhizobium* strains nodulating *Kummerowia* obtained from previous studies and isolated in this study were all covered here. The high genetic diversity of 14 genospecies is shown on the phylogenetic tree based on three taxonomic markers (*SMc*00019, *thrA* and *truA*), which were also previously introduced as taxonomic markers to differentiate the relationship of intraspecific and interspecific genomes among rhizobial strains. The high Hd (0.981) and π (0.116 31) values shown in Tab. 4-4 also indicate the high biodiversity of *Kummerowia* bradyrhizobia.

In addition, the phylogenetic trees of the concatenated sequences of three ecoregion-related genes (*atpD*, *glnII* and *recA*), which are also taxonomic markers, were similar to those of three taxonomic markers, which mirrored the high genetic diversity of *Kummerowia* bradyrhizobia. Nonetheless, the phylogenetic relationship of some individuals was altered, such as many strains embraced in species *Bradyrhizobium* sp. I and *B. elkanii* were separated, and the strains *B. diazoefficiens* CCBAU 33037 and *B. yuanmingense* CCBAU 61035 were located far away from their own groups (Fig. 4-2). For symbiosis-related genes, the genetic diversity was relatively lower and the *Bradyrhizobium* strains belonging to different genospecies were embraced on the phylogenetic trees (Fig. 4-3), which was inferred through the selective communication between legumes and rhizobia. At the same time, the incongruences of the phylogenetic trees revealed by the SH test of single genes compared with three concatenated taxonomic gene phylogeny suggested that the evolutionary direction of each gene was significantly different although they belong to the same genospecies.

Interestingly, many *Bradyrhizobium* strains isolated from WFs, including southern (six provinces) and northern (one province) China, were closely located and mingled together based on three kinds of test genes, which showed a paradox of biogeographical patterns for legume-nodulating *Rhizobium*, *Ensifer*, *Burkholderia* and other species. However, an independent group was formed by the strains isolated from UAs, indicating that the genetic relationships of *Bradyrhizobium* strains were greatly affected by local environmental variables of wild or cultivated *Kummerowia* with human intervention (known as urban segregation) than by geographic isolation, and strains from UAs will be more easily involved in convergent evolution in the future than those from WFs. However, the strain *B. yuanmingense* CCBAU 101063 isolated from UAs closely gathered to the strains from WFs on the phylogenetic tree for all kinds of genes and was inferred as one of the few persistent native rhizobia without radical genetic divergence. Thus, the biodiversity of *Bradyrhizobium* strains being driven to lower levels might be due to the legume *Kummerowia*, which has suffered urban segregation and artificial cultivation by humans.

Infrequent gene flow and recombination among *Kummerowia* *Bradyrhizobium* with urban segregation

The gene flow among rhizobial strains isolated from the roots of *Glycine max*, *Arachis hypogaea*, *Medicago sativa*, *Caragana* and other legumes was frequent for the core genes or symbiosis-related genes, regardless of the genospecies and geographical distribution. However, bradyrhizobial strains of *Kummerowia* isolated from UAs in this study had less genetic communication with the isolates from WFs, evidenced by the lower Nm values between WFs and UAs in southern and northern China. It was also suggested that artificial cultivation and urban segregation could cvitically lower the gene flow frequencies among *Kummerowia* bradyrhizobia for all the test genes. This was also supported by previous studies of the rhizobia associated with *Astragalus*. Moreover, genetic divergence among *Kummerowia* bradyrhizobia was strongly affected by urban segregation, rather than by geographical distribution, due to the high Dxy values between WFs and UAs (Tab. 4-6) in southern and northern China (Tab. 4-7).

In addition to gene flow, an important impetus of evolutionary history is recombination, which has been shown in recent studies to make a contribution to creating diversity of rhizobia comparable with or greater than that of mutation. For example, the higher Rm and r/m (>1.0) values indicate that it was recombination, rather than mutation, that played a vital role in the evolutionary history of the three taxonomic markers and ecoregion-related genes. Nevertheless, all the ρ/θ and r/m values calculated in UAs were far lower than those calculated in WFs, particularly for symbiosis-related genes, thereby indicating a lower frequency of recombination among *Kummerowia* bradyrhizobia in UAs. This observation is consistent with the significantly lower Rm values in UAs. Therefore, recombination and genetic drift events among the strains frequently happened under the stronger natural selection in WFs, but neutral mutation as the critical force occurred in UAs.

Independent evolutionary lineage of *Kummerowia* bradyrhizobia evolved with urban segregation

Rhizobial strains isolated from the root nodules of *Kummerowia* in this study consisted of four lineages (I -IV) for the three kinds of genes [Fig. 4-6 (a), (b), (c)]. For the three taxonomic markers and symbiosis-related genes, lineages I - III were mainly distributed in WFs, although the patterns between them had large differences, while lineage IV, an independent branch, was almost exclusively located in UAs, with the exception of strain CCBAU 101063. Horizontal gene transfer was also rarely observed between *Kummerowia* bradyrhizobia isolated from UAs and WFs in the results of STRUCTURE analyses (Fig. 4-6). This indicates that the evolutionary lineages of *Kummerowia* bradyrhizobia independently evolved for the three taxonomic markers and symbiosis-related genes. Strain CCBAU 101063 might be one of the rhizobial populations which has preserved inherent efficient symbiosis and nitrogen fixation on host plants and strong viability in UA conditions, whereas other strains isolated from UAs as the dominant populations in lineage IV might have had higher efficient symbiosis and nitrogen fixation or stronger viability. By contrast, lineages I - III showed very similar results for the three taxonomic markers and ecoregion-related genes, regardless of the representative strains isolated from southern or northern China (Fig. 4-6). This indicates that an independent lineage of the rhizobia would gradually evolve in the history, primarily due to urban segregation rather than biogeographic separation, which has already been established to be a critical factor related to microbial distribution.

Conclusion

In conclusion, our results confirm that the diversity of rhizobial strains associated with *Kummerowia* has drastically reduced in UA environmental conditions, although the formation of rhizobial communities is generally affected not only by biogeographic separation but also by their host legumes in natural wild environments.

Gene exchange and recombination were frequent within the WFs and UAs separately, but seldom occurred between the two ecoregions based on the analyses of three types of genes. Furthermore, a single evolutionary lineage gradually formed for the *Bradyrhizobium* species of cultivated *Kummerowia* with urban segregation.

References

[1] Li H Y, Li H Q, Li Q H, et al. Multi-species and multi-scale patterns and species associations of sand-fixing plantations [J]. Chinese Journal of Applied Ecology, 2008, 19(4): 741-746.

[2] Xia G M, Kang S Z, Du T S, et al. Transpiration of *Hedysarum scoparium* in arid desert region of Shiyang River basin, Gansu Province[J]. Chinese Journal of Applied Ecology, 2007, 18(6): 1194-1202.

[3] Xiao C W, Jia F P, Zhou G S, et al. Response of photosynthesis, morphology and growth of *Hedysarum mongolicum* seedlings to simulated precipitation change in Maowusu sandland[J]. J. Environ. Sci. (China), 2002, 14(2): 277-283.

[4] Li L, Li X Y, Xu X W, et al. Assimilative branches and leaves of the desert plant *Alhagi sparsifolia* Shap. possesses a different adaptation mechanism to shade[J]. Plant Physiol. Biochem. , 2014, 74: 239-245.

[5] Li H F, Zeng F J, Gui D W, et al. Effects of different disturbance modes on the morphological characteristics and aboveground biomass of *Alhagi sparsifolia* in oasis-desert ecotone[J]. Chinese Journal of Applied Ecology, 2012, 23(1): 23-28.

[6] Chen W M, Zhu W F, C Bontemps, et al. *Mesorhizobium camelthorni* sp. nov. , isolated from *Alhagi sparsifolia*[J]. Int. J. Syst. Evol. Microbiol. , 2011, 61: 574-579.

[7] Chen W M, Zhu W F, Bontemps C, et al. *Mesorhizobium alhagi* sp. nov. , isolated from wild *Alhagi sparsifolia* in north-western China[J]. Int. J. Syst.

Evol. Microbiol. , 2010, 60(Pt 4): 958-962.

[8] Zhao L F, Xu Y J, Ma Z Q, et al. Colonization and plant growth promoting characterization of endophytic *Pseudomonas chlororaphis* strain Zong1 isolated from *Sophora alopecuroides* root nodules[J]. Braz. J. Microbiol. , 2013, 44 (2): 623-31.

[9] Yu H, Song Y, Xi B, et al. Denitrification potential and its correlation to physico-chemical and biological characteristics of saline wetland soils in semi-arid regions[J]. Chemosphere, 2012, 89(11): 1339-1346.

[10] Chen W F, Guan S H, Zhao C T, et al. Different *Mesorhizobium* species associated with *Caragana* carry similar symbiotic genes and have common host ranges[J]. FEMS Microbiol. Lett. , 2008, 283(2): 203-209.

[11] Li M, Li Y, Chen W F, et al. Genetic diversity, community structure and distribution of rhizobia in the root nodules of *Caragana* spp. from arid and semi-arid alkaline deserts, in the north of China[J]. Syst. Appl. Microbiol. , 2012, 35(4): 239-245.

[12] Hubisz M J, Falush D, Stephens M, et al. Inferring weak population structure with the assistance of sample group information[J]. Mol. Ecol. Res. , 2009, 9(5): 1322-1332.

[13] Moukoumi J, Hynes R K, Dumonceaux T J, et al. Characterization and genus identification of rhizobial symbionts from *Caragana arborescens* in western Canada[J]. Can. J. Microbiol. , 2013, 59(6): 399-406.

[14] Jian S Q, Zhao C Y, Fang S M, et al. Characteristics of rainfall interception by *Caragana korshinskii* and *Hippophae rhamnoides* in Loess Plateau of Northwest China [J]. Chinese Journal of Applied Ecology, 2012, 23 (9): 2383-2389.

[15] Zhang W, Hu Y G, Huang G H, et al. Soil microbial diversity of artificial peashrub plantation on North Loess Plateau of China [J]. Acta Microbiol. Sin. , 2007, 47(5): 751-756.

[16] Yan X R, Chen W F, Fu J F, et al. *Mesorhizobium* spp. are the main microsymbionts of *Caragana* spp. grown in Liaoning Province of China[J].

FEMS Microbiol. Lett. , 2007, 271(2): 265-273.

[17] Hanson C A, Fuhrman J A, Horner-Devine M C, et al. Beyond biogeographic patterns: processes shaping the microbial landscape[J]. Nat. Rev. Microbiol. , 2012, 10(7): 497-506.

[18] Xu L, Huber H, During H J, et al. Intraspecific variation of a desert shrub species in phenotypic plasticity in response to sand burial[J]. New Phytol. , 2013, 199(4): 991-1000.

[19] Bhardwaj P K, Kapoor R, Mala D, et al. Braving the attitude of altitude: *Caragana jubata* at work in cold desert of Himalaya[J]. Sci. Rep. , 2013, 3: 1022.

[20] Liu R T, Chai Y Q, Xu K, et al. Variations of ground vegetation and soil properties during the growth process of artificial sand-fixing *Caragana intermedia* plantations in desert steppe [J]. Chinese Journal of Applied Ecology, 2012, 23(11): 2955-2960.

[21] Brigido C, Alexandre A, Oliveira S. Transcriptional analysis of major chaperone genes in salt-tolerant and salt-sensitive mesorhizobia[J]. Microbiol. Res. , 2012, 167(10): 623-629.

[22] Pinto F G, Chueire L M, Vasconcelos A T, et al. Novel genes related to nodulation, secretion systems, and surface structures revealed by a genome draft of *Rhizobium tropici* strain PRF 81[J]. Funct. Integr. Genom. , 2009, 9(2): 263-270.

[23] Wais R J, Keating D H, Long S R. Structure-function analysis of nod factor-induced root hair calcium spiking in *Rhizobium*-legume symbiosis[J]. Plant Physiol. , 2002, 129(1): 211-224.

[24] Estévez J, Soria-Díaz M E, de Córdoba F F, et al. Different and new Nod factors produced by *Rhizobium tropici* CIAT899 following Na^+ stress [J]. FEMS Microbiol. Lett. , 2009, 293(2): 220-231.

[25] López-Lara I M, Geiger O. The nodulation protein NodG shows the enzymatic activity of an 3-oxoacyl-acyl carrier protein reductase[J]. Molecular Plant-Microbe Interactions, 2001, 14(3): 349-357.

[26] Verástegui-Valdés M M, Zhang Y J, Rivera-Orduña F N, et al. Microsymbionts of *Phaseolus vulgaris* in acid and alkaline soils of Mexico[J]. Syst. Appl. Microbiol., 2014, 37(8): 605-612.

[27] Ardley J K, Parker M A, De Meyer S E, et al. *Microvirga lupini* sp. nov., *Microvirga lotononidis* sp. nov. and *Microvirga zambiensis* sp. nov. are alphaproteobacterial root-nodule bacteria that specifically nodulate and fix nitrogen with geographically and taxonomically separate legume hosts[J]. Int. J. Syst. Evol. Microbiol., 2012, 62(Pt 11): 2579-2588.

[28] Kim D H, Kaashyap M, Rathore A, et al. Phylogenetic diversity of *Mesorhizobium* in chickpea[J]. J. Biosci., 2014, 39(3): 513-517.

[29] Li W, Wang C, Wang H, et al. Distribution of atmospheric particulate matter (PM) in rural field, rural village and urban areas of northern China[J]. Environ. Pollut., 2014, 185: 134-140.

[30] Zhang K C, Qu J J, Zu R P, et al. Environmental characteristics of sandstorm of Minqin Oasis in China for recent 50 years[J]. J. Environ. Sci., 2005, 17(5): 857-860.

[31] Terefework Z, Kaijalainen S, Lindström K. AFLP fingerprinting as a tool to study the genetic diversity of *Rhizobium galegae* isolated from *Galega orientalis* and *Galega officinalis*[J]. J. Biotechnol., 2001, 91(2-3): 169-180.

[32] Brigido C, Robledo M, Menéndez E, et al. A ClpB chaperone knockout mutant of *Mesorhizobium ciceri* shows a delay in the root nodulation of chickpea plants[J]. Mol. Plant Microbe Interact., 2012, 25(12): 1594-1604.

[33] Nydam M L, Hoang T A, Shanley K M, et al. Molecular evolution of a polymorphic HSP40-like protein encoded in the histocompatibility locus of an invertebrate chordate[J]. Dev. Comp. Immunol., 2013, 41(2): 128-136.

[34] Botha M, Chiang A N, Needham P G, et al. *Plasmodium falciparum* encodes a single cytosolic type I Hsp40 that functionally interacts with Hsp70 and is upregulated by heat shock[J]. Cell Stress Chaperones, 2011, 16(4): 389-401.

[35] Bhangoo M K, Tzankov S, Fan A C, et al. Multiple 40-kDa heat-shock pro-

tein chaperones function in Tom70-dependent mitochondrial import[J]. Mol. Biol. Cell, 2007, 18(9): 3414-3428.

[36] Hayashi M, Imanaka-Yoshida K, Yoshida T, et al. A crucial role of mitochondrial Hsp40 in preventing dilated cardiomyopathy [J]. Nat. Med., 2006, 12(1): 128-132.

[37] Shim E H, Kim J I, Bang E S, et al. Targeted disruption of hsp70.1 sensitizes to osmotic stress[J]. EMBO Rep., 2002, 3(9): 857-861.

[38] Rico A I, Gironés N, Fresno M, et al. The heat shock proteins, Hsp70 and Hsp83, of *Leishmania infantum* are mitogens for mouse B cells[J]. Cell Stress Chaperones, 2002, 7(4): 339-346.

[39] Piano A, Asirelli C, Caselli F, et al. Hsp70 expression in thermally stressed *Ostrea edulis*, a commercially important oyster in Europe[J]. Cell Stress Chaperones, 2002, 7(3): 250-257.

[40] Zhang Y, Nijbroek G, Sullivan M L, et al. Hsp70 molecular chaperone facilitates endoplasmic reticulum-associated protein degradation of cystic fibrosis transmembrane conductance regulator in yeast[J]. Mol. Biol. Cell, 2001, 12(5): 1303-1314.

[41] Lee J W, Park E, Jeong M S, et al. HslVU ATP-dependent protease utilizes maximally six among twelve threonine active sites during proteolysis[J]. J. Biol. Chem., 2009, 284(48): 33475-33484.

[42] Diouf D, Fall D, Chaintreuil C, et al. Phylogenetic analyses of symbiotic genes and characterization of functional traits of *Mesorhizobium* spp. strains associated with the promiscuous species *Acacia seyal* Del. [J]. J. Appl. Microbiol., 2010, 108(3): 818-830.

[43] Provorov N A, Andronov E E, Onishchuk O P, et al. Genetic structure of the introduced and local populations of *Rhizobioum leguminosarum* in plant-soil systems[J]. Microbiology, 2012, 81(2): 224-232.

[44] Hou B, Li F, Yang X, et al. A small functional intramolecular region of NodD was identified by mutation[J]. Acta Biochim. Biophys. Sin. (Shanghai), 2009, 41(10): 822-830.

[45] Laranjo M, Young J P, Oliveira S. Multilocus sequence analysis reveals multiple symbiovars within *Mesorhizobium* species[J]. Syst. Appl. Microbiol. , 2012, 35(6): 359-367.

[46] Martens M, Dawyndt P, Coopman R, et al. Advantages of multilocus sequence analysis for taxonomic studies: a case study using 10 housekeeping genes in the genus *Ensifer* (including former *Sinorhizobium*) [J]. Int. J. Syst. Evol. Microbiol. , 2008, 58(Pt 1): 200-214.

[47] Vinuesa P, Silva C, Werner D, et al. Population genetics and phylogenetic inference in bacterial molecular systematics: the roles of migration and recombination in *Bradyrhizobium* species cohesion and delineation[J]. Mol. Phylogenet. Evol. , 2005, 34(1): 29-54.

[48] Rivas R, Martens M, De Lajudie P, et al. Multilocus sequence analysis of the genus *Bradyrhizobium*[J]. Syst. Appl. Microbiol. , 2009, 32(2): 101-110.

[49] Elliott G N, Chen W M, Bontemps C, et al. Nodulation of *Cyclopia* spp. (Leguminosae, Papilionoideae) by *Burkholderia tuberum*[J]. Ann. Bot. , 2007, 100(7): 1403-1411.

[50] Sarita S, Sharma P K, Priefer U B, et al. Direct amplification of rhizobial *nodC* sequences from soil total DNA and comparison to *nodC* diversity of root nodule isolates[J]. FEMS Microbiol. Ecol. , 2005, 54(1): 1-11.

[51] Tamura K, Peterson D, Peterson N, et al. MEGA5: molecular evolutionary genetics analysis using maximum likelihood, evolutionary distance, and maximum parsimony methods [J]. Mol. Biol. Evol. , 2011, 28 (10): 2731-279.

[52] Guindon S, Dufayard J F, Lefort V, et al. New algorithms and methods to estimate maximum-likelihood phylogenies: assessing the performance of PhyML 3.0[J]. Syst. Biol. , 2010, 59(3): 307-321.

[53] Posada D, Crandall K A. MODELTEST: testing the model of DNA substitution[J]. Bioinformatics, 1998, 14(9): 817-818.

[54] Huson D H, Bryant D. Application of phylogenetic networks in evolutionary

studies[J]. Mol. Biol. Evol., 2006, 23(2): 254-267.

[55] Librado P, Rozas J. DnaSP v5: a software for comprehensive analysis of DNA polymorphism data[J]. Bioinformatics, 2009, 25(11): 1451-1452.

[56] Nei M, Gojobori T. Simple methods for estimating the numbers of synonymous and nonsynonymous nucleotide substitutions[J]. Mol. Biol. Evol., 1986, 3(5): 418-426.

[57] Didelot X, Falush D. Inference of bacterial microevolution using multilocus sequence data[J]. Genetics, 2007, 175(3): 1251-1266.

[58] Jakobsson M, Rosenberg N A. CLUMPP: a cluster matching and permutation program for dealing with label switching and multimodality in analysis of population structure[J]. Bioinformatics, 2007, 23(14): 1801-1806.

[59] Shimodaira H, Hasegawa M. Multiple comparisons of log-likelihoods with applications to phylogenetic inference[J]. Mol. Biol. Evol., 1999, 16(8): 1114-1116.

[60] Hudson R R, Kaplan N L. Statistical properties of the number of recombination events in the history of a sample of DNA sequences[J]. Genetics, 1985, 111(1): 147-164.

[61] Gao L, Deng X, Wang H, et al. Diversity and resistance of rhizobia isolated from *Caragana intermedia* in Maowusu sandland [J]. Chin. J. Appl. Eco., 2004, 15(1): 44-48.

[62] Guan S H, Chen W F, Wang E T, et al. *Mesorhizobium caraganae* sp. nov., a novel rhizobial species nodulated with *Caragana* spp. in China[J]. Int. J. Syst. Evol. Microbiol., 2008, 58(Pt 11): 2646-2653.

[63] Degefu T, Wolde-meskel E, Frostegard A. Multilocus sequence analyses reveal several unnamed *Mesorhizobium* genospecies nodulating *Acacia* species and *Sesbania* sesban trees in Southern regions of Ethiopia[J]. Syst. Appl. Microbiol., 2011, 34(3): 216-226.

[64] Zhao L, Fan M, Zhang D, et al. Distribution and diversity of rhizobia associated with wild soybean (*Glycine soja* Sieb. & Zucc.) in Northwest China [J]. Syst. Appl. Microbiol., 2014, 37(6): 449-456.

[65] Lu Y L, Chen W F, Han L L, et al. *Rhizobium alkalisoli* sp. nov., isolated from *Caragana intermedia* growing in saline-alkaline soils in the north of China[J]. Int. J. Syst. Evol. Microbiol., 2009, 59(Pt 12): 3006-3011.

[66] Guo H J, Wang E T, Zhang X X, et al. Replicon-dependent differentiation of symbiosis-related genes in *Sinorhizobium* strains nodulating *Glycine max* [J]. Appl. Environ. Microbiol., 2014, 80(4): 1245-1255.

[67] Xu K W, Penttinen P, Chen Y X, et al. Polyphasic characterization of rhizobia isolated from *Leucaena leucocephala* from Panxi, China[J]. World J. Microbiol. Biotechnol., 2013, 29(12): 2303-2315.

[68] Degefu T, Wolde-Meskel E, Liu B, et al. *Mesorhizobium shonense* sp. nov., *Mesorhizobium hawassense* sp. nov. and *Mesorhizobium abyssinicae* sp. nov., isolated from root nodules of different agroforestry legume trees[J]. Int. J. Syst. Evol. Microbiol., 2013, 63(Pt 5): 1746-1753.

[69] Noisangiam R, Teamtisong K, Tittabutr P, et al. Genetic diversity, symbiotic evolution, and proposed infection process of *Bradyrhizobium* strains isolated from root nodules of *Aeschynomene americana* L. in Thailand[J]. Appl. Environ. Microbiol., 2012, 78(17): 6236-6250.

[70] Delamuta J R, Ribeiro R A, Menna P, et al. Multilocus sequence analysis (MLSA) of *Bradyrhizobium* strains: revealing high diversity of tropical diazotrophic symbiotic bacteria[J]. Braz. J. Microbiol., 2012, 43(2): 698-710.

[71] Aujoulat F, Jumas-Bilak E, Masnou A, et al. Multilocus sequence-based analysis delineates a clonal population of *Agrobacterium* (*Rhizobium*) *radiobacter* (*Agrobacterium tumefaciens*) of human origin[J]. J. Bacteriol., 2011, 193(10): 2608-2618.

[72] Zhang X X, Guo H J, Wang R, et al. Genetic divergence of *Bradyrhizobium* strains nodulating soybeans as revealed by multilocus sequence analysis of genes inside and outside the symbiosis island[J]. Appl. Environ. Microbiol., 2014, 80(10): 3181-90.

[73] Tang J, Bromfield E S, Rodrigue N, et al. Microevolution of symbiotic *Bra-*

dyrhizobium populations associated with soybeans in east North America[J]. Ecol. Evol., 2012, 2(12): 2943-2961.

[74] Vinuesa P, Rojas-Jiménez K, Contreras-Moreira B, et al. Multilocus sequence analysis for assessment of the biogeography and evolutionary genetics of four *Bradyrhizobium* species that nodulate soybeans on the asiatic continent [J]. Appl. Environ. Microbiol., 2008, 74(22): 6987-6996.

[75] Silva C, Kan F L, Martínez-Romero E. Population genetic structure of *Sinorhizobium meliloti* and *S. medicae* isolated from nodules of *Medicago* spp. in Mexico[J]. FEMS Microbiol. Ecol., 2007, 60(3): 477-489.

[76] Torres-Leguizamon M, Mathieu J, Decaëns T, et al. Genetic structure of earthworm populations at a regional scale: Inferences from mitochondrial and microsatellite molecular markers in *Aporrectodea icterica* (Savigny 1826) [J]. PLoS ONE, 2014, 9(7): e101597.

[77] Willi Y, Griffin P, Van Buskirk J. Drift load in populations of small size and low density[J]. Heredity (Edinb), 2013, 110(3): 296-302.

[78] Petren K, Grant P R, Grant B R, et al. Multilocus genotypes from Charles Darwin's finches: biodiversity lost since the voyage of the *Beagle*[J]. Philos. Trans. R. Soc. Lond. B Biol. Sci., 2010, 365(1543): 1009-1018.

[79] Isaenko O A, Karr T L, Feder M E. Hsp70 and thermal pretreatment mitigate developmental damage caused by mitotic poisons in *Drosophila*[J]. Cell Stress Chaperones, 2002, 7(3): 297-308.

[80] Feder M E, Hofmann G E. Heat-shock proteins, molecular chaperones, and the stress response: evolutionary and ecological physiology[J]. Ann. Rev. Physiol., 1999, 61: 243-282.

[81] Charlesworth B, Morgan M T, Charlesworth D. The effect of deleterious mutations on neutral molecular variation [J]. Genetics, 1993, 134(4): 1289-1303.

[82] Kimura M. DNA and the neutral theory[J]. Philos. Trans. R. Soc. Lond. B Biol. Sci., 1986, 312(1154): 343-354.

[83] Kimura M. The neutral theory of molecular evolution[J]. Sci. Am., 1979,

241(5): 98-108,102,108.

[84] Han L L, Wang E T, Han T X, et al. Unique community structure and biogeography of soybean rhizobia in the saline-alkaline soils of Xinjiang, China [J]. Plant and Soil, 2009, 324(1): 291-305.

[85] Zhang Y M, Li Y, Chen W F, et al. Biodiversity and biogeography of rhizobia associated with soybean plants grown in the North China Plain[J]. Appl. Environ. Microbiol., 2011, 77(18): 6331-6342.

[86] Zhang J J, Liu T Y, Chen W F, et al. *Mesorhizobium muleiense* sp. nov., nodulating with *Cicer arietinum* L. [J]. Int. J. Syst. Evol. Microbiol., 2012, 62: 2737-2742.

[87] Verástegui-Valdés M M, Zhang Y J, Rivera-Orduña F N, et al. Microsymbionts of *Phaseolus vulgaris* in acid and alkaline soils of Mexico[J]. Syst. Appl. Microbiol., 2014, 37(8): 605-612.

[88] Wu L J, Wang H Q, Wang E T, et al. Genetic diversity of nodulating and non-nodulating rhizobia associated with wild soybean (*Glycine soja* Sieb. & Zucc.) in different ecoregions of China[J]. FEMS Microbiol. Ecol., 2011, 76(3): 439-450.

[89] Mutch L A, Young J P. Diversity and specificity of *Rhizobium leguminosarum* biovar *viciae* on wild and cultivated legumes[J]. Mol. Ecol., 2004, 13(8): 2435-2444.

[90] Armas-Capote N, Pérez-Yépez J, Martínez-Hidalgo P, et al. Core and symbiotic genes reveal nine *Mesorhizobium* genospecies and three symbiotic lineages among the rhizobia nodulating *Cicer canariense* in its natural habitat (La Palma, Canary Islands) [J]. Syst. Appl. Microbiol., 2014, 37(2): 140-148.

[91] Laranjo M, Alexandre A, Rivas R, et al. Chickpea rhizobia symbiosis genes are highly conserved across multiple *Mesorhizobium* species[J]. FEMS Microbiol. Ecol., 2008, 66(2): 391-400.

[92] Shao B M, Xu W, Dai H, et al. A study on the immune receptors for polysaccharides from the roots of *Astragalus membranaceus*, a Chinese medicinal

herb[J]. Biochem. Biophy. Res. Commun., 2004, 320(4): 1103-1111.

[93] Lin L Z, He X G, Lindenmaier M, et al. Liquid chromatography-electrospray ionization mass spectrometry study of the flavonoids of the roots of *Astragalus mongholicus* and *A. membranaceus*[J]. J. Chromato. A, 2000, 876: 87-95.

[94] Zhao C T, Wang E T, Chen W F, et al. Diverse genomic species and evidences of symbiotic gene lateral transfer detected among the rhizobia associated with *Astragalus* species grown in the temperate regions of China[J]. FEMS Microbiol. Lett., 2008, 286(2): 263-273.

[95] Wei G H, Tan Z Y, Zhu M E, et al. Characterization of rhizobia isolated from legume species within the genera *Astragalus* and *Lespedeza* grown in the Loess Plateau of China and description of *Rhizobium loessense* sp. nov[J]. Int. J. Syst. Evol. Microbiol., 2003, 53(Pt 5): 1575-1583.

[96] Ren D W, Wang E T, Chen W F, et al. *Rhizobium herbae* sp. nov. and *Rhizobium giardinii*-related bacteria, minor microsymbionts of various wild legumes in China[J]. Int. J. Syst. Evol. Microbiol., 2011, 61(Pt 8): 1912-1920.

[97] Laguerre G, Van Berkum P, Amarger N, et al. Genetic diversity of rhizobial symbionts isolated from legume species within the genera *Astragalus*, *Oxytropis*, and *Onobrychis*[J]. Appl. Environ. Microbiol., 1997, 63(12): 4748-4758.

[98] Li Q F, Zhang X P, Zou L, et al. Horizontal gene transfer and recombination shape mesorhizobial populations in the gene center of the host plants *Astragalus luteolus* and *Astragalus ernestii* in Sichuan, China[J]. FEMS Microbiol. Ecol., 2009, 70(2): 71-79.

[99] Gao J, Terefework Z, Chen W, et al. Genetic diversity of rhizobia isolated from *Astragalus adsurgens* growing in different geographical regions of China [J]. J. Biotechnol., 2001, 91(2-3): 155-168.

[100] Zhang X X, Turner S L, Guo X W, et al. The common nodulation genes of *Astragalus sinicus* rhizobia are conserved despite chromosomal diversity[J].

Appl. Environ. Microbiol. , 2000, 66(7): 2988-2995.

[101] Chen W M, Sun L L, Lu J J, et al. Diverse nodule bacteria were associated with *Astragalus* species in arid region of northwestern China[J]. J. Basic Microbiol. , 2013, 55(1): 121-128.

[102] Gnat S, Wójcik M, Wdowiak-Wróbel S, et al. Phenotypic characterization of *Astragalus glycyphyllos* symbionts and their phylogeny based on the 16S rDNA sequences and RFLP of 16S rRNA gene[J]. Ant. Van Leeuwen. , 2014, 105(6): 1033-1048.

[103] Guerrouj K , Pérez-Valera E, Chahboune R, et al. Identification of the rhizobial symbiont of *Astragalus glombiformis* in Eastern Morocco as *Mesorhizobium camelthorni*[J]. Ant. Van Leeuwen. , 2013, 104(2): 187-198.

[104] Dong Y M, Tang D Y, Zhang N, et al. Phytochemicals and biological studies of plants in genus *Hedysarum*[J]. Chem. Cent. J. , 2013, 7: 124

[105] Houba V J G, Novozamsky I, Huybregts A W M, et al. Comparison of soil extractions by 0.01 M $CaCl_2$, by EUF and by some conventional extraction procedures[J]. Plant and Soil, 1986, 96(3): 433-437.

[106] Simonis A D. Effect of temperature on extraction of phosphorus and potassium from soils by various extracting solutions [J]. Commun. Soil Sci. Plant Ana. , 1996, 27(3-4): 665-684.

[107] Terefework Z, Kaijalainen S, Lindström K. AFLP fingerprinting as a tool to study the genetic diversity of *Rhizobium galegae* isolated from *Galega orientalis* and *Galega officinalis*[J]. J. Biotech. , 2001, 91(2): 169-180.

[108] Tan Z Y, Wang E T, Peng G X, et al. Characterization of bacteria isolated from wild legumes in the north-western regions of China[J]. Int. J. Syst. Bacteriol. , 1999, 49(Pt4): 1457-1469.

[109] Wang E T, Van Berkum P, Sui X H, et al. Diversity of rhizobia associated with *Amorpha fruticosa* isolated from Chinese soils and description of *Mesorhizobium amorphae* sp. nov [J]. Int. J. Syst. Bacteriol. , 1999, 49 (Pt1): 51-65.

[110] Normand P, Cournoyer B, Simonet P, et al. Analysis of a ribosomal RNA

operon in the actinomycete *Frankia*[J]. Gene, 1992, 111(1): 119-124.

[111] Laguerre G, Mavingui P, Allard M R, et al. Typing of rhizobia by PCR DNA fingerprinting and PCR-restriction fragment length polymorphism analysis of chromosomal and symbiotic gene regions: application to *Rhizobium leguminosarum* and its different biovars[J]. Appl. Environ. Microbiol., 1996, 62(6): 2029-2036.

[112] Germano M G, Menna P, Mostasso F L, et al. RFLP analysis of the rRNA operon of a Brazilian collection of bradyrhizobial strains from 33 legume species[J]. Int. J. Syst. Evol. Microbiol., 2006, 56(1): 217-229.

[113] Vinuesa P, Silva C, Werner D, et al. Population genetics and phylogenetic inference in bacterial molecular systematics: the roles of migration and recombination in *Bradyrhizobium* species cohesion and delineation[J]. Mol. Phylogenet. Evol., 2005, 34(1): 29-54.

[114] Laguerre G, Nour S M, Macheret V, et al. Classification of rhizobia based on *nodC* and *nifH* gene analysis reveals a close phylogenetic relationship among *Phaseolus vulgaris* symbionts[J]. Microbiology, 2001, 147(Pt4): 981-993.

[115] Tamura K, Peterson D, Peterson N, et al. MEGA5: molecular evolutionary genetics analysis using maximum likelihood, evolutionary distance, and maximum parsimony methods[J]. Mol. Biol. Evol., 2011, 28(10): 2731-2739.

[116] Ji Z J, Yan H, Cui Q G, et al. Genetic divergence and gene flow among *Mesorhizobium* strains nodulating the shrub legume *Caragana*[J]. Syst. Appl. Microbiol., 2015, 38(3): 176-183.

[117] Ormeño-Orrillo E, Servín-Garcidueñas L E, Rogel M A, et al. Taxonomy of rhizobia and agrobacteria from the *Rhizobiaceae* family in light of genomics[J]. Syst. Appl. Microbiol., 2015, 38(4): 287-291.

[118] Rao C R. The use and interpretation of principal component analysis in applied research[J]. Sankhya, A, 1964, 26(4): 329-358.

[119] Hill T C J, Walsh K A, Harris J A, et al. Using ecological diversity mea-

sures with bacterial communities[J]. FEMS Microbiol. Ecol. , 2003, 43 (1): 1-11.

[120] Ibáñez F, Angelini J, Taurian T, et al. Endophytic occupation of peanut root nodules by opportunistic *Gammaproteobacteria*[J]. Syst. Appl. Microbiol. , 2009, 32(1): 49-55.

[121] Li L, Sinkko H, Montonen L, et al. Biogeography of symbiotic and other endophytic bacteria isolated from medicinal *Glycyrrhiza* species in China [J]. FEMS Microbiol. Ecol. , 2012, 79(1): 46-68.

[122] Zhang J J, Yu T, Lou K, et al. Genotypic alteration and competitive nodulation of *Mesorhizobium muleiense* against exotic chickpea rhizobia in alkaline soils[J]. Syst. Appl. Microbiol. , 2014, 37(7): 520-524.

[123] Donnarumma F, Bazzicalupo M, Blažinkov M, et al. Biogeography of *Sinorhizobium meliloti* nodulating alfalfa in different Croatian regions[J]. Res. Microbiol. , 2014, 165(7): 508-16.

[124] Sullivan J T, Patrick H N, Lowther W L, et al. Nodulating strains of *Rhizobium loti* arise through chromosomal symbiotic gene transfer in the environment[J]. Proc. Natl. Acad. Sci. USA, 1995, 92(19): 8985-8989.

[125] Yan H, Ji Z J, Jiao Y S, et al. Genetic diversity and distribution of rhizobia associated with the medicinal legumes *Astragalus* spp. and *Hedysarum polybotrys* in agricultural soils[J]. Syst. Appl. Microbio. , 2016, 39(2): 141-149.

[126] Zhao C T, Wang E T, Chen W F, et al. Diverse genomic species and evidences of symbiotic gene lateral transfer detected among the rhizobia associated with *Astragalus* species grown in the temperate regions of China[J]. FEMS Microbiol. Lett. , 2008, 286(2): 263-73.

[127] Li M, Li Y, Chen W F, et al. Genetic diversity, community structure and distribution of rhizobia in the root nodules of *Caragana* spp. from arid and semi-arid alkaline deserts, in the north of China[J]. Sys. Appl. Microbiol. , 2012, 35(4): 239-245.

[128] Lu Y L, Chen W F, Wang E T, et al. Genetic diversity and biogeography

of rhizobia associated with *Caragana* species in three ecological regions of China[J]. Syst. Appl. Microbiol. , 2009, 32(5): 351-361.

[129] Ji Z J, Yan H, Cui Q G, et al. Competition between rhizobia under different environmental conditions affects the nodulation of a legume[J]. Syst. Appl. Microbiol. , 2017, 40(2): 114-119.

[130] Hirsch A M, Lum M R, Downie J A. What makes the rhizobia-legume symbiosis so special? [J]. Plant Physiology, 2001, 127(4): 1484-1492.

[131] Okazaki S, Kaneko T, Sato S, et al. Hijacking of leguminous nodulation signaling by the rhizobial type III secretion system[J]. Proc. Natl. Acad. Sci. USA, 2013, 110(42): 17131-17136.

[132] Tian C F, Zhou Y J, Zhang Y M, et al. Comparative genomics of rhizobia nodulating soybean suggests extensive recruitment of lineage-specific genes in adaptations [J]. Proc. Natl. Acad. Sci. USA, 2012, 109 (22): 8629-8634.

[133] Sugawara M, Epstein B, Badgley B D, et al. Comparative genomics of the core and accessory genomes of 48 *Sinorhizobium* strains comprising five genospecies[J]. Genome Biol. , 2013, 14(2): R17.

[134] Kamst E, Pilling J, Raamsdonk L M, et al. Rhizobium nodulation protein NodC is an important determinant of chitin oligosaccharide chain length in Nod factor biosynthesis[J]. J. Bacteriol. , 1997, 179(7): 2103-2108.

[135] Guan S H, Chen W F, Wang E T, et al. *Mesorhizobium caraganae* sp. nov. , a novel rhizobial species nodulated with *Caragana* spp. in China [J]. Int. J. Syst. Evol. Microbiol. , 2008, 58(Pt11): 2646-2653.

[136] Luo R, Liu B, Xie Y, et al. SOAPdenovo2: an empirically improved memory-efficient short-read *de novo* assembler [J]. Gigascience, 2012, 27 (1):18.

[137] Delcher A, Bratke K, Powers E, et al. Identifying bacterial genes and endosymbiont DNA with Glimmer [J]. Bioinformatics, 2007, 23 (6): 673-679.

[138] Salzberg S, Delcher A, Kasif S, et al. Microbial gene identification using

interpolated Markov models[J]. Nucleic Acids Research, 1998, 26(2): 544-548.

[139] Zhao Y, Wu J Y, Yang J H, et al. PGAP: pan-genomes analysis pipeline [J]. Bioinformatics, 2012, 28(3): 416-418.

[140] Eddy S R. Accelerated profile HMM searches[J]. PLoS Comput. Biol., 2011, 7(10): e1002195.

[141] Larkin M A, Blackshields G, Brown N P, et al. Clustal W and Clustal X version 2.0[J]. Bioinformatics, 2007, 23(21): 2947-2948.

[142] Guindon S, Dufayard J F, Lefort V, et al. New algorithms and methods to estimate maximum-likelihood phylogenies: assessing the performance of PhyML 3.0[J]. Syst. Biol., 2010, 59(3): 307-321.

[143] Huson D H, Bryant D. Application of phylogenetic networks in evolutionary studies[J]. Mol. Biol. Evol., 2006, 23(2): 254-267.

[144] Tamura K, Stecher G, Peterson D, et al. MEGA6: molecular evolutionary genetics analysis version 6.0 [J]. Mol. Biol. Evol., 2013, 30(12): 2725-2729.

[145] Wolk C P, Cai Y, Panoff J. Use of a transposon with luciferase as a reporter to identify environmentally responsive genes in a cyanobacterium[J]. Pro. Natl. Acad. Sci. USA, 1991, 88(12): 5355-5359.

[146] Jiao Y S, Liu Y H, Yan H, et al. Rhizobial diversity and nodulation characteristics of the extremely promiscuous legume *Sophora flavescens* [J]. Mol. Plant-Microbe Interact., 2015, 28(12): 1338-1352.

[147] Bloemberg G V, Kamst E, Harteveld M, et al. A central domain of Rhizobium NodE protein mediates host specificity by determining the hydrophobicity of fatty acyl moieties of nodulation factors[J]. Mol. Microbiol., 1995, 16(6): 1123-1136.

[148] Demont N, Debellé F, Aurelle H, et al. Role of the *Rhizobium meliloti* nodF and nodE genes in the biosynthesis of lipo-oligosaccharidic nodulation factors[J]. J. Biol. Chem., 1993, 268(27): 20134-20142.

[149] Quinto C, Wijfjes A H M, Bloemberg G V, et al. Bacterial nodulation pro-

tein NodZ is a chitin oligosaccharide fucosyltransferase which can also recognize related substrates of animal origin[J]. Proc. Natl. Acad. Sci. USA, 1997, 94(9): 4336-4341.

[150] Stacey G, Luka S, Sanjuan J, et al. *nodZ*, a unique host-specific nodulation gene, is involved in the fucosylation of the lipooligosaccharide nodulation signal of *Bradyrhizobium japonicum*[J]. J. Bacteriol., 1994, 176 (3): 620-633.

[151] Quesada-Vincens D, Fellay R, Nasim T, et al. *Rhizobium* sp. strain NGR234 NodZ protein is a fucosyltransferase[J]. J. Bacteriol., 1997, 179 (16): 5087-5093.

[152] Rodpothong P, Sullivan S J, Songsrirote K, et al. Nodulation gene mutants of *Mesorhizobium loti* R7A—*nodZ* and *nolL* mutants have host-specific phenotypes on *Lotus* spp. [J]. Mol. Plant-Microbe Interact., 2009, 22(12): 1546-1554.

[153] Schubert K R, Evans H J. Hydrogen evolution: A major factor affecting the efficiency of nitrogen fixation in nodulated symbionts [J]. Proc. Natl. Acad. Sci. USA, 1976, 73(4): 1207-1211.

[154] Brito B, Martínez M, Fernández D, et al. Hydrogenase genes from *Rhizobium leguminosarum* bv. *viciae* are controlled by the nitrogen fixation regulatory protein nifA [J]. Proc. Natl. Acad. Sci USA, 1997, 94 (12): 6019-6024.

[155] Ruiz-Argüeso T, Palacios J M, Imperial J. Regulation of the hydrogenase system in *Rhizobium leguminosarum*[J]. Plant and Soil, 2001, 230(1): 49-57.

[156] Murillo J, Villa A, Chamber M, et al. Occurrence of H_2-uptake hydrogenases in *Bradyrhizobium* sp. (*Lupinus*) and their expression in nodules of *Lupinus* spp. and *Ornithopus compressus* [J]. Plant physiol., 1989, 89 (1): 78-85.

[157] Baginsky C, Brito B, Imperial J, et al. Symbiotic hydrogenase activity in *Bradyrhizobium* sp. (*Vigna*) increases nitrogen content in *Vigna* unguicula-

ta plants[J]. Appl. Environ. Microbiol. , 2005, 71(11): 7536-7538.

[158] Ruiz-Argüeso T, Hanus J, Evans H J. Hydrogen production and uptake by pea nodules as affected by strains of *Rhizobium leguminosarum* [J]. Archives of Microbiology, 1978, 116(2): 113-118.

[159] Fernández D, Toffanin A, Palacios J M, et al. Hydrogenase genes are uncommon and highly conserved in *Rhizobium leguminosarum* bv. *viciae*[J]. FEMS Microbiol. Lett. , 2005, 253(1): 83-88.

[160] Van Berkum, P. , R. B. Navarro, and A. Vargas. Classification of the uptake hydrogenase-positive (Hup$^+$) bean rhizobia as *Rhizobium tropici*[J]. Applied and environmental microbiology, 1994, 60(2): 554-561.

[161] Van Berkum P, Navarro R B, Vargas A A. Diversity and Evolution of Hydrogenase Systems in Rhizobia[J]. Appl. Environ. Microbiol. , 2002, 60(2): 554-561.

[162] Scheu A K, Economou A, Hong G F, et al. Secretion of the *Rhizobium leguminosarum* nodulation protein NodO by haemolysin-type systems [J]. Mol. Microbiol. , 1992, 6(2): 231-238.

[163] Sutton J, Peart J, Dean G, et al. Analysis of the C-terminal secretion signal of the *Rhizobium leguminosarum* nodulation protein NodO; a potential system for the secretion of heterologous proteins during nodule invasion[J]. Mol. Plant-Microbe. Interacti. , 1996, 9(8): 671-680.

[164] Fauvart M, Michiels J. Rhizobial secreted proteins as determinants of host specificity in the rhizobium-legume symbiosis[J]. FEMS Microbio. Lett. , 2008, 285(1): 1-9.

[165] Downie J A, Surin B P. Either of two nod gene loci can complement the nodulation defect of a nod deletion mutant of *Rhizobium leguminosarum* bv. *viciae*[J]. Mol. Gen. Genet. , 1990, 222(1): 81-86.

[166] Van Rhijn P, Luyten E, Vlassak K, et al. Isolation and characterization of a pSym locus of *Rhizobium* sp. BR816 that extends nodulation ability of narrow host range *Phaseolus vulgaris* symbionts to *Leucaena leucocephala*[J]. Mol. Plant-Microbe Interact. , 1996, 9(1): 74-77.

[167] Vlassak K M, Luyten E, Verreth C, et al. The *Rhizobium* sp. BR816 *nodO* gene can function as a determinant for nodulation of *Leucaena leucocephala*, *Phaseolus vulgaris*, and *Trifolium repens* by a diversity of *Rhizobium* spp. [J]. Mol. Plant-Microbe Interact. , 1998, 11(5): 383-392.

[168] Economou A, Davies A E, Johnston A W B, et al. The *Rhizobium leguminosarum* biovar *viciae nodO* gene can enable a *nodE* mutant of *Rhizobium leguminosarum* biovar *trifolii* to nodulate vetch [J]. Microbiology, 1994, 140(9): 2341-2347.

[169] Economou A, Hamilton W D O, Johnston A W B, et al. The *Rhizobium* nodulation gene *nodO* encodes a Ca^{2+}-binding protein that is exported without N-terminal cleavage and is homologous to haemolysin and related proteins[J]. The EMBO Journal, 1990, 9(2): 349-354.

[170] Sutton J M, Lea E J, Downie J A. The nodulation-signaling protein NodO from *Rhizobium leguminosarum* biovar *viciae* forms ion channels in membranes[J]. Proc. Natl. Acad. Sci. USA, 1994, 91(21): 9990-9994.

[171] Österman J, Mousavi S A, Koskinen P, et al. Genomic features separating ten strains of *Neorhizobium galegae* with different symbiotic phenotypes[J]. BMC Genomics, 2015, 16: 348.

[172] Salmond G P C. Secretion of extracellular virulence factors by plant pathogenic bacteria[J]. Annual Review of Phytopathology, 1994, 32: 181-200.

[173] Lenders M H H, Weidtkamp-Peters S, Kleinschrodt D, et al. Directionality of substrate translocation of the hemolysin A Type I secretion system[J]. Scientific Reports, 2015, 5(1):12470.

[174] Thomas S, Holland I B, Schmitt L. The type 1 secretion pathway-the hemolysin system and beyond[J]. Biochim. Biophys. Acta, 2014, 1843(8): 1629-1641.

[175] Tseng T, Tyler B M, Setubal J C. Protein secretion systems in bacterial-host associations, and their description in the Gene Ontology[J]. BMC Microbiol. , 2009, 9(1): 1-9.

[176] Mazurier S, Lemunier M, Hartmann A, et al. Conservation of type III se-

cretion system genes in *Bradyrhizobium* isolated from soybean[J]. FEMS Microbiol. Lett., 2006, 259(2): 317-325.

[177] Okazaki S, Okabe S, Higashi M, et al. Identification and functional analysis of type III effector proteins in *Mesorhizobium loti*[J]. Mol. Plant-Microbe Interact., 2010, 23(2): 223-234.

[178] Kim W S, Krishnan H B. A nopA deletion mutant of *Sinorhizobium fredii* USDA257, a soybean symbiont, is impaired in nodulation[J]. Curr. Microbiol., 2013, 68(2): 239-246.

[179] Okazaki S, Tittabutr P, Teulet A, et al. Rhizobium-legume symbiosis in the absence of Nod factors: two possible scenarios with or without the T3SS [J]. ISME J., 2016, 10(1): 64-74.

[180] Viprey V, Del Greco A, Golinowski W, et al. Symbiotic implications of type III protein secretion machinery in *Rhizobium*[J]. Mol. Microbiol., 1998, 28(6): 1381-1389.

[181] Schmeisser C, Liesegang H, Krysciak D, et al. *Rhizobium* sp. strain NGR234 possesses a remarkable number of secretion systems[J]. Appl. Environ. Microbiol., 2009, 75(12): 4035-4045.

[182] Kaneko T, Nakamura Y M, Sato S, et al. Complete genomic sequence of nitrogen-fixing symbiotic bacterium *Bradyrhizobium japonicum* USDA110 [J]. DNA research, 2002, 9(6): 189-197.

[183] Tampakaki A P. Commonalities and differences of T3SSs in rhizobia and plant pathogenic bacteria[J]. Front. Plant Sci., 2014, 5: 114.

[184] González V, Santamaría R I, Bustos P, et al. The partitioned *Rhizobium etli* genome: Genetic and metabolic redundancy in seven interacting replicons [J]. Proc. Natl. Acad. Sci. USA, 2006, 103(10): 3834-3839.

[185] Pueppke S G, Broughton W J. *Rhizobium* sp. strain NGR234 and *R. fredii* USDA257 share exceptionally broad, nested host ranges[J]. Mol. Plant-Microbe Interact. USA, 1999, 12(4): 293-318.